Formal Verification of Structurally Complex Multipliers

Alireza Mahzoon • Daniel Große • Rolf Drechsler

Formal Verification of Structurally Complex Multipliers

Alireza Mahzoon
University of Bremen
Bremen, Germany

Rolf Drechsler
University of Bremen/DFKI
Bremen, Germany

Daniel Große
Institute for Complex Systems
Johannes Kepler University of Linz
Linz, Austria

ISBN 978-3-031-24573-2 ISBN 978-3-031-24571-8 (eBook)
https://doi.org/10.1007/978-3-031-24571-8

This Springer imprint is published by the registered company Springer Nature Switzerland AG
The registered company address is: Gewerbestrasse 11, 6330 Cham, Switzerland

To Tina,
Marie
and
Yuna

Preface

Back in 1970, an Intel 4004 processor had 2250 transistors. It could only support a limited number of instructions, and it was working at a very low frequency. However, digital circuits nowadays are much larger, sometimes even consisting of billions of transistors. Moreover, they are usually designed based on sophisticated algorithms, leading to fast but complex architectures. The big size and the high complexity of modern digital circuits make them extremely error-prone during different design phases. Consequently, formal verification is an important task to ensure the correctness of a digital circuit.

Formal verification of arithmetic circuits is one of the most challenging problems in the verification community. Despite the success of verification methods based on *Binary Decision Diagrams* (BDDs) and *Boolean Satisfiability* (SAT) in proving the correctness of adders, they totally fail when it comes to the verification of multipliers. In the last six years, the word-level verification methods based on *Symbolic Computer Algebra* (SCA) achieved many successes in proving the correctness of structurally simple multipliers. The proposed techniques can verify a very large multiplier in a few seconds. However, they either totally fail or support a limited set of benchmarks when it comes to verifying structurally complex multipliers.

This book addresses the challenging tasks of verifying and debugging structurally complex multipliers. In the area of verification, it first investigates the challenges of SCA-based verification when it comes to proving the correctness of multipliers. Then, it proposes three techniques, i.e., vanishing monomials removal, reverse engineering, and dynamic backward rewriting, to improve and extend SCA. As a result, a wide variety of multipliers, including highly complex and optimized industrial benchmarks, can be verified. In the area of debugging, it proposes a complete debugging flow, including bug localization and fixing, to find the location of bugs in structurally complex multipliers and make corrections.

Bremen, Germany — Alireza Mahzoon
Linz, Austria — Daniel Große
Bremen, Germany — Rolf Drechsler

Acknowledgements

We would like to particularly appreciate all those who have contributed to the results included in this book. Our special thanks go to Christoph Scholl and Alexander Konrad for their investments of time and great ideas, which were important for this book. Many thanks to Mehran Goli for his helpful feedback and inspiring discussions. We would like to express our appreciation to all of the colleagues in the Group of Computer Architecture at the University of Bremen, the Institute for Complex Systems at the Johannes Kepler University Linz, and the Cyber-Physical Systems group at the German Research Center for Artificial Intelligence for their support.

Bremen, Germany
Linz, Austria
Bremen, Germany
October 2022

Alireza Mahzoon
Daniel Große
Rolf Drechsler

Contents

Chapter 1
Introduction

With the invention of the transistor back in 1947, the cornerstone for the digital revolution was laid. As a fundamental building block, the transistor triggered the development of digital circuits. The mass production of digital circuits revolutionized the field of electronics, finally leading to computers, embedded systems, and the Internet. Hence, the impact of digital hardware on society, as well as the economy, was and is tremendous. Over the last decades, the enormous growth in the complexity of integrated circuits continues as expected. As modern electronic devices are getting more and more ubiquitous, the fundamental issue of functional correctness becomes more important than ever. This is evidenced by many publicly known examples of electronic failures with disastrous consequences. This includes, e.g., the Intel Pentium bug in 1994 [6, 28], the New York blackout in 2003 [64], and a design flaw in Intel's Sandy Bridge chipset in 2011 [18].

Such costly mistakes can only be prevented by applying rigorous verification to the circuits before they get to production [20, 21]. A lot of effort has been put into developing efficient verification techniques by both academic and industrial research. Only recently, the industry has recognized the great importance of formal verification (see, e.g., functional safety standards such as ISO 26262 [88]). Hence, in the last few years, this research area has become increasingly active. Essentially, formal verification aims to prove in a mathematical sense that an implementation is correct with respect to its specification. Formal verification is an essential task in each phase of the design flow to ensure the correctness of an implementation.

An overview of the typical top-down design flow is presented in Fig. 1.1. The system specification defines the functionality and is usually the starting point for the design. The register transfer model is constructed by a designer or a high-level synthesis tool based on the system specification. At this level, the system behavior is described in terms of registers and their data flow with its operations. Logic synthesis tools transfer the model to a gate-level description, consisting of logic gates and flip-flops. Finally, the logic gates are mapped into a circuit-level description, consisting of transistors, interconnects, and other physical cells,

A. Mahzoon et al., *Formal Verification of Structurally Complex Multipliers*,
https://doi.org/10.1007/978-3-031-24571-8_1

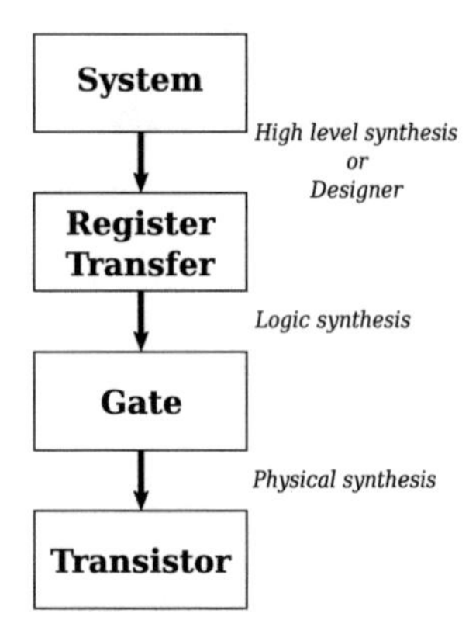

Fig. 1.1 Top-down design flow

which create the final chip implementation. There is always a risk of incorrect transformations by designers or synthesis tools when moving between different levels of abstraction. As a result, bugs might appear in each phase of the design, leading to a faulty circuit description. The fabrication and production of faulty designs cause a catastrophe, resulting in huge financial loss and endangering lives. It is thus critical to ensure the correctness of a circuit description at each level of abstraction.

Several formal verification methods have been proposed to prove the correctness of a circuit description. These methods can be categorized into three groups:

- **Equivalence checking**: The goal of formal equivalence checking is to prove that a circuit description is functionally equivalent to a specification (golden model). The specification is usually a correct description in the same level or a higher level of abstraction. For example, assuming that gate-level description *A* is correct, gate-level description *B* will be correct if *A* and *B* are functionally equivalent. As another example, the gate-level description and the high-level system specification have to be equivalent; otherwise, the gate-level description is faulty. Equivalence checking is widely employed in the automated formal verification of both combinational and sequential digital circuits.
- **Model checking**: The aim of model checking is to ensure that a property holds for a circuit description in a specific level of abstraction. This includes safety properties (nothing incorrect ever occurs) and liveness properties (something correct eventually occurs). In model checking, the circuit description is captured as a transition system, specifying its behavior in different states. Furthermore, the property is expressed in the form of a temporal logic formula, and a model checker is used to check whether the property is violated or not. Model checking is widely employed in the automated formal verification of sequential digital circuits.
- **Theorem proving**: The goal of theorem proving in verification is to prove that a circuit description satisfies its specification by mathematical reasoning. The description and the specification are expressed as formulas in a formal

logic. Then, the required relationship between them (logical equivalence or logical implication) is described as a theorem to be proven within the context of a proof calculus. A proof system, consisting of axioms and interface rules (e.g., simplification, rewriting, and induction), is used to achieve this goal.

Formal verification of arithmetic circuits is one of the most popular and challenging topics in the verification community. Arithmetic circuits are extensively used in many systems, e.g., for signal processing and cryptography, as well as for upcoming AI solutions employing machine learning and deep learning. They also constitute a big part of an *Arithmetic Logic Unit* (ALU), which is the computational heart of a *Central Processing Unit* (CPU). The top-level design flow in Fig. 1.1 is also used for the implementation of arithmetic circuits. The high-level specification is usually a mathematical expression, determining the function of an arithmetic circuit based on its primary inputs and outputs. For example, assuming A and B are two n-bit inputs and S is an $(n + 1)$-bit output, the expression $S = A + B$ describes an integer adder in the highest level of abstraction. The high-level specification is transformed into the register transfer level by a designer or an arithmetic generator tool. Finally, the circuit is synthesized into gate-level and then transistor-level descriptions. Formal verification of arithmetic circuits at the register transfer level (where the hierarchical information is available) and the gate level (where no hierarchical information is at hand) is the focus of many research works.

Integer multipliers are among the most frequently used arithmetic circuits in a large variety of applications. Most of these applications require very large multipliers supporting a wide range of integer numbers. Furthermore, the multiplier architectures also vary based on the design goals of different applications. Several multiplication algorithms have been developed to satisfy the community demands for fast, area-efficient, and low-power designs or make a trade-off between several design parameters. Employing these algorithms usually results in the generation of very complex architectures. Formal verification of huge and structurally complex multipliers is on the one hand necessary to ensure the correctness of the final design. On the other hand, it is a big challenge, where most of the existing formal methods completely fail.

In the last 30 years, several formal verification methods, based on equivalence checking and theorem proving techniques, have been proposed to ensure the correctness of arithmetic circuits. Although these methods accomplished big successes in many domains, they suffer from serious limitations when it comes to verifying integer multipliers:

- Equivalence checking methods using *Binary Decision Diagrams* (BDDs) [10, 27] or *Boolean Satisfiability* (SAT) [19, 35] ensure the correctness by proving that an integer multiplier is equivalent to a correct multiplier description. However, they are not scalable and only work for very small benchmarks.
- Equivalence checking methods using *Binary Moment Diagrams* (*BMDs and K*BMDs) [23, 39] ensure the correctness by proving that an integer multiplier and its high-level specification are equivalent. These methods are scalable for

structurally simple multipliers. However, they fail to verify structurally complex multipliers, since they face a memory blow-up when the multiplier size increases.
- Theorem proving methods [81, 82, 85] use a library of rewrite rules to prove that a multiplier description can correctly implement its specification. However, for new multiplier architectures, the database of rewrite rules has to be updated. Otherwise, the theorem prover cannot complete the proof. Thus, they need high manual effort. Moreover, these methods require hierarchical information, which is available in the register transfer description. As a result, they cannot be used to verify the gate-level description of a multiplier.

Recently, *Symbolic Computer Algebra* (SCA) verification methods have shown very good results in proving the correctness of large but structurally simple integer multipliers [31, 71, 93, 94]. They have also been employed in verifying floating point multipliers [76], finite field multipliers [37, 66], and dividers [43, 77, 78, 91, 92] as well as debugging faulty integer multipliers [33, 34, 74], and finite field multipliers [36, 67–69]. The SCA-based verification is categorized as an equivalence checking approach, consisting of three main steps:

1. The function of a multiplier is represented based on its primary inputs and outputs as a *Specification Polynomial* (SP).
2. The logic gates (or nodes of an *AND-Inverter Graph* (AIG)) are captured as a set of polynomials P_G.
3. Gröbner basis theory is used to prove the membership of SP in the ideal generated by P_G.

The third step consists of the step-wise division of SP by P_G (or equivalently, substitution of variables in SP with P_G), known as *backward rewriting* and eventually the evaluation of the remainder. If the remainder is zero, the multiplier is correct. Otherwise, it is buggy. In fact, SCA-based verification proves that the gate-level description of a multiplier and its high-level specification (represented in the form of SP) are equivalent. **In this book, we make two major contributions for SCA which can be assigned to two areas: verification and debugging.**

Verification SCA-based methods have successfully verified structurally simple multipliers (i.e., multipliers whose second and third stages are only made of half-adders and full-adders). However, the verification of structurally complex multipliers (i.e., multipliers whose second and third stages are not fully made of half-adders and full-adders) is a big challenge for these methods as an explosion happens in the number of monomials during backward rewriting. This creates major barriers for the industrial use since industrial multipliers are usually structurally complex.

In general, SCA-based verification faces two critical challenges when it comes to the verification of structurally complex multipliers:

- Generation of redundant monomials, known as *vanishing monomials*, causes a blow-up in the size of intermediate polynomials during backward rewriting. These monomials are generated during the verification and reduced to zero

after many steps. However, the huge number of vanishing monomials before cancellation causes an explosion in computations.
- There are usually many possible orderings for the substitution of gate/node polynomials in the intermediate polynomial during backward rewriting. Some of these orderings increase the size of intermediate polynomial drastically, while the others keep the size small.

Recently, there have been some attempts to overcome the first challenge [75] or totally avoid both challenges by design alterations [41]. However, the application of these methods is limited since they only partially support the verification of structurally complex multipliers. Especially, they are not robust when it comes to the verification of optimized multipliers. As a result, there is currently a high demand for an SCA-based verification method that overcomes both challenges and successfully verifies structurally complex multipliers, including optimized architectures used in industry. **In this book, we develop innovative theories, algorithms, and data structures for SCA to verify structurally complex multipliers (Contribution I).**

Debugging If a formal verification method proves that an arithmetic circuit is buggy; then, localizing and fixing bugs become the major subsequent tasks. Although the verification of arithmetic circuits, including integer multipliers, has received considerable attention, the number of research works on the debugging of arithmetic circuits is very limited. Recently, researchers have attempted to employ SCA in debugging integer multipliers [33, 34, 74]. They take advantage of the non-zero remainder at the end of backward rewriting to localize and fix bugs. However, these methods are not applicable in many cases:

- They fail to debug faulty architectures when the bug causes the propagation of vanishing monomials to the remainder. As a result, bug localization with the help of the remainder becomes impossible.
- They fail to debug faulty multipliers when the bug is not close to the primary inputs. The bug adds several new monomials to the process of backward rewriting, which eventually leads to an explosion in the number of monomials.

Thus, the SCA-based debugging methods can be only used to localize and fix bugs in very limited cases. The lack of a complete automated debugging flow for multipliers is a serious problem since manual debugging is very time-consuming and costly. **In this book, we come up with a complete debugging flow based on SCA and SAT to localize and fix bugs in both structurally simple and complex multipliers, regardless of the location of bugs (Contribution II).**

1.1 Overview

This book makes two major contributions for SCA in the areas of verification and debugging:

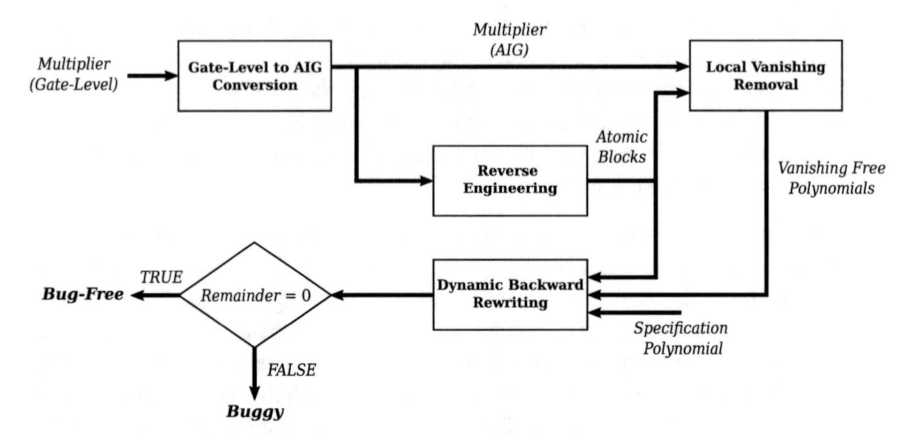

Fig. 1.2 Our proposed SCA-based verification

Contribution I Figure 1.2 depicts the proposed method to verify structurally complex multipliers, which is the first contribution of this book. The method extends the basic SCA-based verification by introducing three techniques:

- **Reverse engineering:** Atomic blocks, i.e., half-adders, full-adders, and compressors are identified in the AIG representation by a dedicated reverse engineering technique. These blocks facilitate the detection of vanishing monomials and significantly speed up global backward rewriting.
- **Local vanishing monomials removal:** Vanishing monomials, which are one of the main reasons for the monomials explosion during backward rewriting, are detected and removed locally. As a result, the global backward rewriting is performed vanishing-free.
- **Dynamic backward rewriting:** The order of polynomials substitutions is set dynamically during global backward rewriting. Moreover, the bad substitutions, increasing the size of intermediate polynomials dramatically, are restored. As a result, it becomes possible to control the size of intermediate polynomials.

The proposed method, including the three aforementioned techniques, is implemented in the SCA-based verifier REVSCA-2.0. The tool receives a gate-level integer multiplier, containing no hierarchical information, and proves its correctness. REVSCA-2.0 supports the verification of various structurally complex multipliers, including optimized and industrial designs.

Contribution II Debugging structurally complex multipliers is the second contribution of this book. The proposed method covers the complete debugging flow, consisting of verification, localization, and fixing. Each task is carried out by a combination of SCA and SAT in order to maximize the performance and avoid the problems of pure SCA-based debugging. Figure 1.3 presents our proposed debugging flow consisting of three main tasks (i.e., verification, localization, and

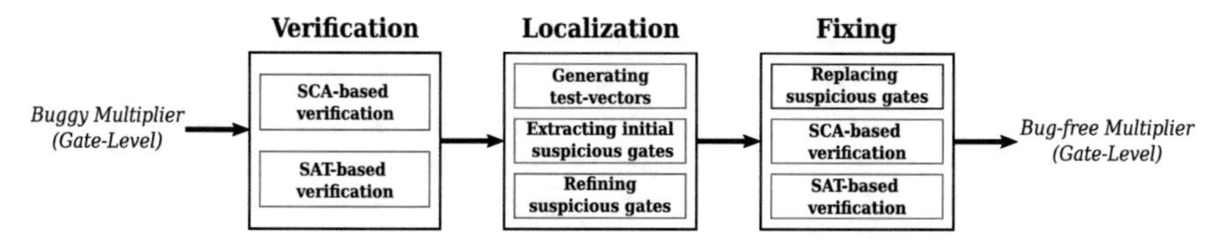

Fig. 1.3 Our proposed debugging flow

fixing) and several sub-tasks. The sub-tasks in the blue boxes are performed by SCA, while the sub-tasks in the red boxes are carried out by SAT. The debugging method can localize and fix bugs in a wide variety of gate-level structurally complex multipliers, regardless of the location of bugs.

1.2 Outline

This book consists of nine chapters, including the current introductory chapter. Chapter 2 presents the necessary background and the state-of-the-art. This chapter first introduces the different ways of circuit modeling; then, it gives insight into integer multiplier structures. Subsequently, formal approaches for multiplier verification are discussed, and their advantages and disadvantages are highlighted. Finally, formal verification using SCA is explained, and the state-of-the-art SCA-based methods are reviewed. The main parts of this book are structured in the following chapters.

- **Contribution I**
  - **Chapter 3** provides a detailed description of structurally simple and complex multipliers. It presents several experimental results to showcase the performance of SCA-based verification when it comes to proving the correctness of different architectures. This chapter discusses the success of SCA-based methods in verifying structurally simple multipliers and their failure in verifying structurally complex multipliers.
  - **Chapter 4** presents a novel technique to locally remove vanishing monomials and avoid an explosion, caused by vanishing monomials, during global backward rewriting. The concept of vanishing monomials is first introduced by an example in this chapter. Then, the observations from the example are generalized, and the basic theory of vanishing monomials is given. Furthermore, the connection between vanishing monomials and multiplier architectures is investigated. Finally, an algorithm to detect vanishing monomials and remove them in structurally complex multipliers is proposed. The proposed technique in this chapter has been published in [50, 53].
  - **Chapter 5** presents a reverse engineering technique to identify atomic blocks in multipliers. The concept of atomic blocks and their importance in the

verification of structurally complex multipliers are first explained in this chapter. Particularly, the role of atomic blocks in vanishing monomials detection is illustrated. Finally, a dedicated reverse engineering technique is proposed to identify three types of atomic blocks, i.e., half-adders, full-adders, and (4:2) compressors. The proposed technique in this chapter has been published in [51, 53].

  - **Chapter 6** presents a dynamic backward rewriting technique to control the size of intermediate polynomials. The challenges of backward rewriting for optimized multipliers are first explained in the chapter. Then, a dynamic backward rewriting is proposed to overcome the challenges by sorting substitution candidates during backward rewriting and restoring bad substitutions. The proposed technique in this chapter has been published in [54].
  - **Chapter 7** presents our SCA-based verifier REVSCA-2.0. REVSCA-2.0 takes advantage of the three introduced techniques in the previous chapters to extend the basic SCA-based verification and prove the correctness of structurally complex multipliers. The details of the REVSCA-2.0 implementation are given in this chapter. Finally, the verification results for a wide variety of structurally complex multipliers, including clean and dirty optimized architectures, are reported and discussed. Parts of this chapter have been published in [52, 53].

- **Contribution II**

  - **Chapter 8** presents our complete debugging flow based on SCA and SAT. After introducing the fault model, the limitations of the pure SCA-based debugging methods are discussed. Then, each phase of our debugging flow, including verification, localization, and fixing, is detailed. The advantage and disadvantages of using SCA and SAT for each phase are explained in this chapter. Finally, the experimental results for several multiplier architectures are reported and discussed. The proposed method in this chapter has been published in [49].

This book is concluded in Chap. 9, which includes a brief discussion on potential research avenues for future work in the area of formal verification and debugging of arithmetic circuits.

Chapter 2
Background

To keep this work self-contained, this chapter provides the basics of multiplier structures as well as the state-of-the-art formal verification techniques proposed to verify them. Moreover, the theoretical background of SCA and its application in the verification of arithmetic circuits are explained, which is the main focus of this book.

This chapter first looks at circuit modeling and reviews the different ways of representing a digital circuit. Then, the structure of one of the most dominant arithmetic units, i.e., integer multiplier, is presented. Subsequently, the state-of-the-art formal verification methods and their efficiency in the verification of arithmetic circuits, particularly integer multipliers, are reviewed.

Finally, the last part of this chapter introduces SCA and the theory of Gröbner basis. It also clarifies how SCA can be employed for the verification of arithmetic circuits, particularly integer multipliers.

2.1 Circuit Modeling

A digital circuit is an electronic circuit that can take on only a finite number of states. This is contrasted with analog circuits whose voltages or other quantities vary in a continuous manner. Binary (two-state) digital circuits are the most common form of digital circuits. The two possible states of a binary circuit are represented by the binary digits, or bits, false (0) and true (1). Binary digital circuits always perform a Boolean function on their input bits. A *Boolean Function* f is a rule (or a formula or an algorithm) that takes a specific number of bits and generates new bits, i.e., $f : \{0, 1\}^k \rightarrow \{0, 1\}^m$, where k and m are non-negative integers.

In the gate-level description of a digital circuit, Boolean functions are implemented by logic gates. The basic logic gates are connected by a network of wires to generate the correct values on the primary outputs with respect to the

A. Mahzoon et al., *Formal Verification of Structurally Complex Multipliers*,
https://doi.org/10.1007/978-3-031-24571-8_2

function. The gate-level description of a digital circuit can be converted to other structural representations, e.g., AIG, *Majority-Inverter Graph* (MIG), and *XOR-AND Graph* (XAG) [83]. Among them, the AIG representation is widely used in different applications such as area and delay optimization, cell mapping, and formal verification.

2.1.1 Gate-Level Netlist

A logic gate is a model of computation or physical electronic device implementing a basic symmetric Boolean function. The functions and symbols of the most common logic gates are shown in Table 2.1 for inputs A and B.

In order to obtain the gate-level description, the circuit is modeled using only logic gates. Figure 2.1a shows the gate-level description of a 2×2 unsigned multiplier. The basic logic gates, including AND and XOR gates, are connected to implement the Boolean function of a multiplier. A gate-level description is usually generated after synthesizing a register transfer description [12, 61, 73]. It is later

Table 2.1 Logic gates symbols and functions

Type	Boolean algebra	Logic gate symbol	Function
BUF	A	A	1 when the input is 1
NOT	$\overline{A}$	A	1 when the input is 0
AND	$A \wedge B$	A B	1 when all inputs are 1
OR	$A \vee B$	A B	1 when any input is 1
NAND	$\overline{A \wedge B}$	A B	1 when it is not the case that all inputs are 1
NOR	$\overline{A \vee B}$	A B	1 when none of the inputs are 1
XOR	$A \oplus B$	A B	1 when one of the inputs is 1 and the other one is 0
XNOR	$A \odot B$	A B	1 when both inputs are 0 or 1

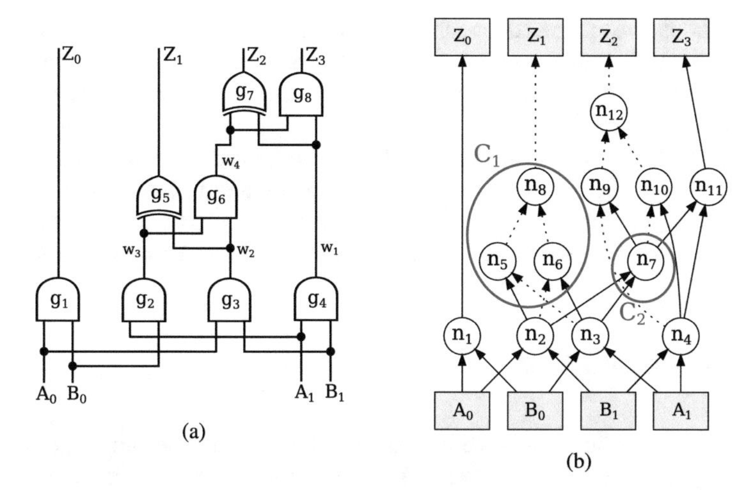

Fig. 2.1 2×2 unsigned multiplier. (**a**) Gate-level representation. (**b**) AIG representation

synthesized into the transistor-level description, which is eventually used in the manufacturing process.

In many verification and optimization problems, the digital circuits are available in the form of the gate-level netlist, describing the types of gates and the connections between them. A gate-level netlist can be converted into other structural representations to facilitate verification and optimization tasks.

2.1.2 AND-Inverter Graph

Definition 2.1 An AIG is a directed acyclic graph composed of two-input AND gates and inverters (NOT gates) with the following properties:

- A node has either zero or two incoming edges.
- A node with no incoming edge is a primary input.
- A node with two incoming edges is an AND gate.
- A complemented edge indicates the negation of a signal.

Figure 2.1b shows the AIG representation of a 2×2 unsigned multiplier. The dashed lines in Fig. 2.1b indicate the complemented edges. A gate-level netlist can be easily converted to the AIG representation. AIGs and particularity the *cut* concept are widely used in logic synthesis since it helps for optimization [60, 90].

Definition 2.2 A *cut* of a node n is a set of nodes C, called leaves, such that (i) every path from n to a primary input must visit at least one node in C and (ii) every node in C must be included in at least one of these paths.

In this book, we are interested in the cone created based on the cut C. The inputs of the cone are C nodes, and the output is n, while it contains all the converging paths from inputs to n. We use the term cut to refer to a cut's cone in the remaining of this book.

In Fig. 2.1b, $C_1 = \{n_5, n_6, n_8\}$ and $C_2 = \{n_7\}$ are cuts for the nodes n_8 and n_7, respectively. The nodes n_2 and n_3 have output edges to both cuts C_1 and C_2; thus, n_2 and n_3 are inputs of C_1 and C_2. Cuts on an AIG can be computed using *cut enumeration* [59, 62], which we later use in this book for reverse engineering.

2.2 Integer Multiplier

Nowadays, arithmetic circuits are inseparable parts of many designs. These digital circuits are getting even more attention due to the high demands for computationally intensive applications (e.g., signal processing and cryptography) as well as upcoming AI architectures (e.g., for machine learning and deep learning). Integer multipliers are among the most frequently used arithmetic circuits. Designers have proposed a variety of multiplier architectures to satisfy the industry's needs for high-speed, low-power, and low-area designs or make a trade-off between several design parameters. These multipliers are usually highly parallel and structurally complex, which makes their verification a challenge for formal methods. This section reviews the structure of integer multipliers in detail.

2.2.1 Structure

Figure 2.2 shows the general structure of an integer multiplier consisting of three stages: *Partial Product Generator* (PPG), *Partial Product Accumulator* (PPA), and *Final Stage Adder* (FSA). The PPG stage generates partial products from the multiplier and the multiplicand inputs. Then, the PPA stage reduces the partial products by multi-operand adders and computes their sum. Eventually, the sum is converted to the corresponding binary output at the FSA [44, 63, 95].

Several algorithms have been proposed to implement each stage of an integer multiplier. The architectures generated by these algorithms have some pros and cons in terms of design parameters, e.g., area, delay, power, and the number of wiring tracks. Designers can choose between different algorithms to achieve the design goals, e.g., minimizing the chip area. For example, the Booth PPG [7] generates fewer partial products compared to the Simple PPG; thus, it reduces the overall area of the multipliers with long operands. However, it has a higher design and logic complexity. As another example, the Wallace tree [87] and the balanced delay tree [96] are two well-known algorithms for implementing the PPA stage. The Wallace tree guarantees the lowest overall delay, but it has the largest number of wiring tracks. On the other hand, the balanced delay tree requires the smallest

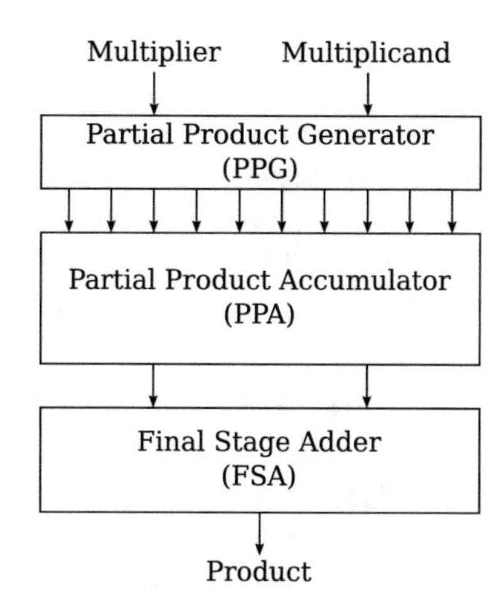

Fig. 2.2 General multiplier structure

Table 2.2 Multiplier architectures and abbreviations

Stage 1 (PPG)	Stage 2 (PPA)	Stage 3 (FSA)
Simple PPG (**SP**)	Array (**AR**)	Ripple carry adder (**RC**)
Booth PPG (**BP**)	Overturned-stairs tree (**OS**)	Conditional sum adder (**CU**)
	Dadda tree (**DT**)	Ladner–Fischer adder (**LF**)
	Wallace tree (**WT**)	Carry look-ahead adder (**CL**)
	Balanced delay tree (**BD**)	Kogge–Stone adder (**KS**)
	Compressor tree (**CT**)	Brent–Kung adder (**BK**)
		Block CL (**BL**)
		Ripple-block CL (**RL**)

number of wiring tracks but suffers from the highest overall delay compared to the other algorithms. As the last example, ripple carry adder and carry look-ahead adder are two algorithms for the FSA implementation. The ripple carry adder has the smallest area, but it suffers from a large delay. In contrast, the carry look-ahead adder has a much smaller delay, but it occupies more area.

In the rest of this book, we use the notation $\boldsymbol{\alpha \circ \beta \circ \gamma}$ to refer to a multiplier consisting of the stages: PPG α, PPA β, and FSA γ. Table 2.2 shows the most common architectures and also their abbreviations for the three stages of a multiplier. We use these abbreviations throughout the book to refer to the architectures.

2.3 Formal Verification of Multipliers

Formal verification and validation are two techniques to check the correctness of a digital circuit against its specification. In validation, it is accomplished through simulation; however, exhaustive simulation for big designs is generally infeasible.

On the other hand, formal verification takes advantage of rigorous mathematical reasoning to prove that a design meets all or parts of its specification [20, 21].

After the famous Pentium bug back in 1994 [65], a lot of effort has been put into the development of formal verification methods for arithmetic circuits. In this section, we review these methods and highlight their advantages and disadvantages, particularly when it comes to verifying integer multipliers.

2.3.1 Equivalence Checking Using BDDs

Definition 2.3 A BDD is a directed, acyclic graph. Each node of the graph has two outgoing edges associated with the values of the variables 0 and 1. A BDD contains two terminal nodes (leaves) that are associated with the values of the function 0 or 1.

Definition 2.4 An *Ordered Binary Decision Diagram* (OBDD) is a BDD, where different variables appear in the same order in each path from the root to a leaf. An OBDD is called a *Reduced Ordered Binary Decision Diagram* (ROBDD) if it has a minimum number of nodes for a given variable order.

In order to formally verify a circuit, we need to have the BDD representation of the outputs. *Symbolic simulation* helps us to obtain the BDD for each primary output. During a simulation, an input pattern is applied to a circuit, and the resulting output values are checked to see whether they match the expected values. On the other hand, symbolic simulation verifies a set of scalar tests (which usually cover the whole input space) with a single symbolic test. Symbolic simulation using BDDs is done by generating corresponding BDDs for the input signals. Then, starting from the primary inputs, the BDD for the output of each logic gate (or building block) is obtained. This process continues until we reach the primary outputs. Finally, the output BDDs are evaluated to see whether they match the BDDs of a specification (golden circuit). An ROBDD is a canonical representation [8]. Thus, if the output BDDs and the BDDs of the golden circuit are equivalent, the correctness is ensured.

BDDs are very efficient in representing addition and subtraction functions, i.e., the size of output BDDs changes linearly with respect to the number of input bits. Moreover, it has been proven that the time complexity for the BDD-based verification of adders and subtractors is polynomial [22]. However, when it comes to the verification of integer multipliers, BDD-based verification can only prove the correctness of small circuits since the size of output BDDs, and subsequently, the verification run-time, grows exponentially with respect to the number of input bits [9]. Figure 2.3 depicts the total size of output BDDs for the multipliers with different sizes. Assuming A and B are inputs of multipliers, we use the variable ordering A_0, B_0, A_1, B_1, …, A_{n-1}, B_{n-1}. The size of BDDs grows exponentially, and subsequently, the machine runs out of memory for the multipliers bigger than 12×12. Please note that choosing other variables' orderings changes the BDD sizes; however, the exponential growth in the BDD sizes happens for all orderings.

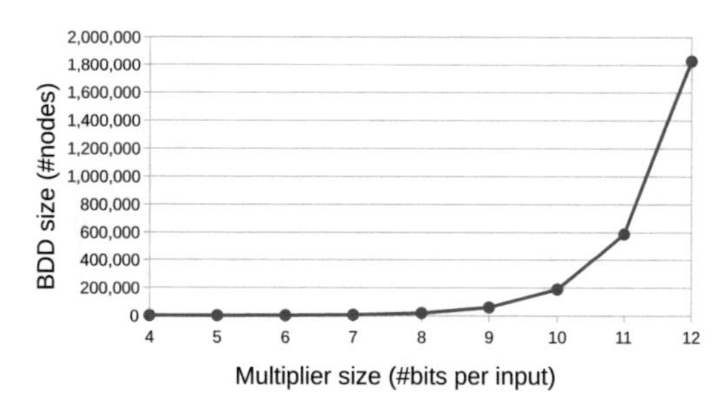

Fig. 2.3 Total size of the output BDDs for the multipliers with different sizes

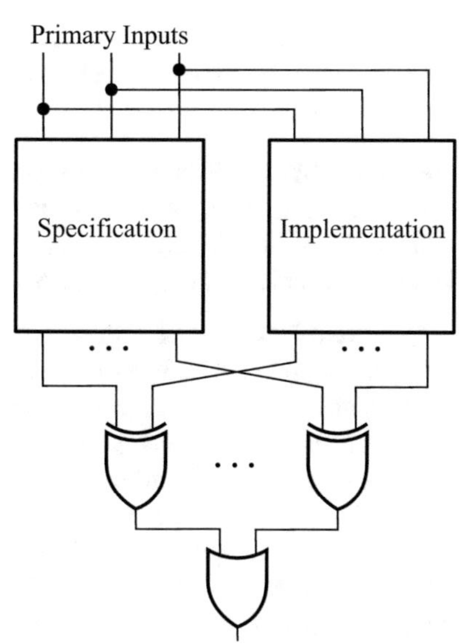

Fig. 2.4 Miter for SAT-based equivalence checking

2.3.2 Equivalence Checking Using SAT

In the SAT-based equivalence checking, the goal is to prove that the implementation and the specification (golden circuit) are equivalent using Boolean satisfiability. It is usually done by creating a miter. A *miter* is a circuit constructed by (1) connecting the corresponding inputs of two circuits, (2) adding 2-input XOR gates on top of corresponding outputs, and (3) connecting the outputs of XOR gates to an OR gate. A miter has only one output which is the output of the OR gate. If the output is always 0 for all possible input values, the two circuits are equivalent. Figure 2.4 shows a miter constructed for the SAT-based verification of two circuits, i.e., specification and implementation.

In order to prove that the output of a miter is always 0, the circuit is transformed into *Conjunctive Normal Form* (CNF), and then it is proven to be unsatisfiable using a SAT-solver.

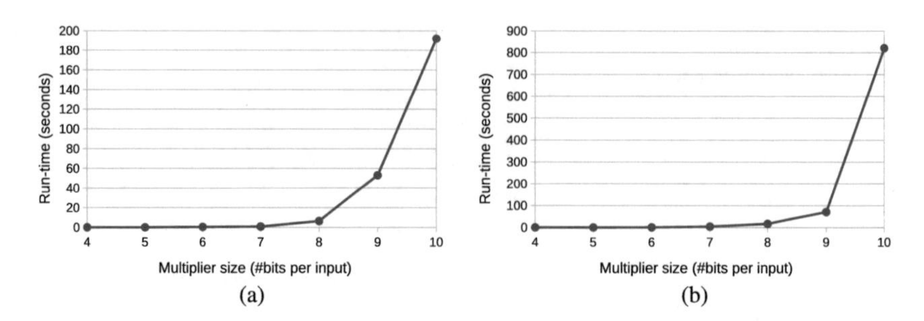

Fig. 2.5 Run-times of CEC for two types of multipliers with different sizes (the multipliers have been checked against correct $SP \circ AR \circ RC$ multipliers). (**a**) $SP \circ WT \circ CL$. (**b**) $BP \circ DT \circ LF$

Definition 2.5 Let V be the set of Boolean variables. Then, the expression b^ϵ with $b \in V$ and $\epsilon \in \{0, 1\}$ is called a *literal*, where $b^0 \equiv \bar{b}$ and $b^1 \equiv b$. A *clause* is a disjunction of literals (e.g., $c = \overline{b_1} \vee b_2 \vee b_3 = (b_1^0, b_2^1, b_3^1)$ is a clause c with the Boolean variables b_1, b_2, and b_3). A CNF is a conjunction of clauses.

The transformation of a miter into the CNF is usually done using the Tseitin transformation [19, 84]. Several techniques have been proposed to speed up the SAT-based equivalence checking. For example, structural hashing [46], identification of intermediate equivalent points [45], and circuit rewriting [59] improved the verification run-time, significantly.

The *Combinational Equivalence Checking* (CEC) using SAT has reported very good results for adders and circuits with structural similarities, e.g., checking the equivalence of a circuit before and after optimization. However, it is not a successful approach when it comes to verifying integer multipliers. It takes a huge amount of time for a SAT-solver to prove the unsatisfiability of the miter's CNF. Therefore, SAT-based equivalence checking cannot prove the correctness of large multipliers in practice.

Figure 2.5a, b show the run-times of CEC (`cec` command in abc [1]) for the $SP \circ WT \circ CL$ and $BP \circ DT \circ LF$ multipliers with different sizes, respectively. An $SP \circ AR \circ RC$ multiplier has been used as the specification. We assume that the circuit specification is always correct. Then, it has been proven that the two aforementioned multipliers are equivalent to the specification. It is evident in Fig. 2.5a, b that the run-times of verifying both multipliers grow exponentially. Consequently, CEC cannot be used to verify multipliers bigger than 10×10 in practice.

2.3.3 *Binary Moment Diagram*

Definition 2.6 A BMD is a directed acyclic graph that represents a function from $\{0, 1\}^n$ to the set of integers, i.e., $\mathbb{B}^n \to \mathbb{Z}$. Each node of a BMD is either a variable node or a constant node. Each constant node is labeled by a distinct integer. Each

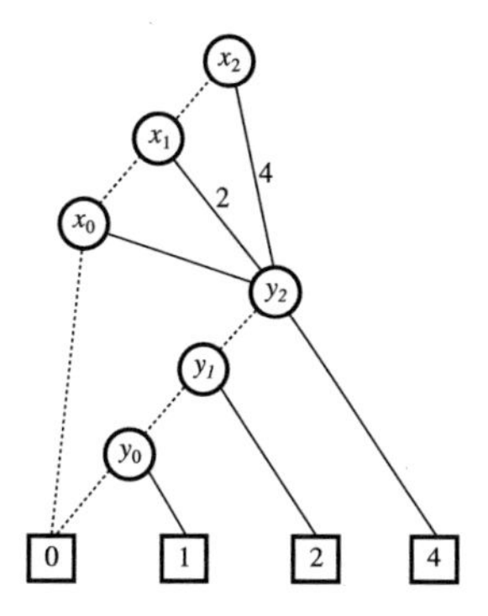

Fig. 2.6 *BMD representation of a 3-bit multiplication function

variable node has two types of edges called 0-edges and 1-edges. The construction of a BMD is done using the moment decomposition:

$$f = f_{\bar{x}_i} + x_i \cdot f_{\dot{x}_i} \qquad (f_{\dot{x}_i} = f_{x_i} - f_{\bar{x}_i}), \tag{2.1}$$

where f_{x_i} and $f_{\bar{x}_i}$ are the functions resulted from the substitution of x_i with 1 and 0, respectively. The BMD representing a function f is constructed recursively so that its root labeled by x is linked via its 0-edge (1-edge) to the root of the BMD representing $f_{\bar{x}_i}$ ($f_{\dot{x}_i}$).

The size of a BMD can be reduced using common factors in the constant moment ($f_{\bar{x}_i}$) and linear moment ($f_{\dot{x}_i}$) in Eq. (2.1). These factors are extracted and placed as so-called *edge-weights* on the incoming edge to the nodes. The new representation is called *Multiplicative Binary Moment Diagram* (*BMD). A *BMD with root vertex v labeled with x and weights w, w_0, and w_1 on the incoming edge to v, on the low-edge of v, and on the high-edge of v, respectively, represents the function $w \cdot f_v$ defined by

$$w \cdot f_v = w \cdot (w_0 \cdot f_{low(v)} + x \cdot w_1 \cdot f_{high(v)}). \tag{2.2}$$

The edge-weights on the low- and high-edge of any non-terminal vertex must have the greatest common divisor 1. Moreover, weight 0 appears only as a terminal value, and if the low-edge of a node points to this terminal node, the weight of the high-edge is 1 [23, 24]. Figure 2.6 shows the *BMD of a 3-bit multiplication function, i.e., $X \times Y = (4x_2 + 2x_1 + x_0) \times (4y_2 + 2y_1 + y_0)$.

The formal verification based on *BMD is one of the first word-level methods successfully used to prove the correctness of some multiplier architectures. This method consists of four main steps:

1. The *BMD of the function which converts the bit-level representation of m outputs to the word-level representation is constructed, i.e., $F = \sum_{i=0}^{m-1} 2^i \cdot z_i$ for unsigned integers and $F = -2^{m-1} + \sum_{i=0}^{m-2} 2^i \cdot z_i$ for signed integers.

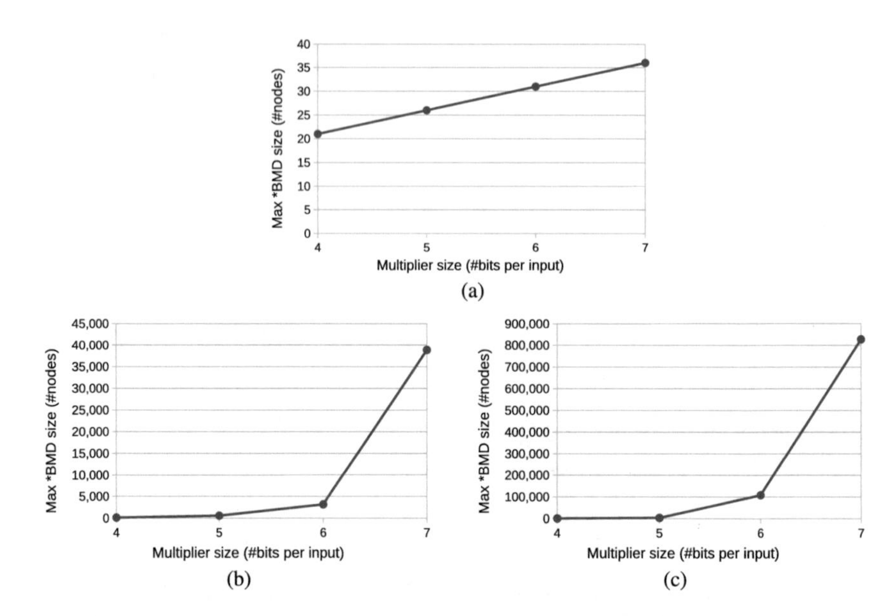

Fig. 2.7 Maximum size of *BMD during backward construction. (**a**) $SP \circ AR \circ RC$. (**b**) $SP \circ WT \circ CL$. (**c**) $BP \circ DT \circ LF$

2. The word-level functions of logic gates are captured as *BMDs, e.g., the word-level function of an OR gate is $G = x + y - xy$.
3. Starting from primary outputs, variable v in the *BMD representation of F, which is the output of a logic gate, is substituted by the *BMD of the corresponding gate function. Then, the *BMD of F is reconstructed. The substitution process is performed for all logic gates in the circuit.
4. It is checked whether the *BMD F represents the function of the circuit, e.g., for a 3-bit multiplier, the final *BMD should match the *BMD in Fig. 2.6.

The word-level formal verification based on *BMD reports good results for the structurally simple multipliers [11, 38], i.e., multipliers whose PPA and FSA stages are only made of half-adders and full-adders. However, when it comes to the verification of structurally complex multipliers, they run out of memory due to the huge size of intermediate *BMDs. Figure 2.7 shows the maximum size of *BMD during backward construction for three types of multipliers with different sizes. The variables are ordered based on the reverse topological order of the circuits. The maximum size of *BMD changes slightly for the $SP \circ AR \circ RC$ multipliers, which have a trivial structure (see Fig. 2.7a). However, the size grows exponentially for the $SP \circ WT \circ CL$ and $BP \circ DT \circ LF$ multipliers (see Fig. 2.7b, c), which are structurally complex, i.e., the third stages are not only made of half-adders and full-adders but also extra gates.

2.4 Term Rewriting

Theorem proving techniques are not usually automated and require considerable manual effort. As a result, their application has been very limited in verifying arithmetic circuits. Recently, the formal verification methods, known as *term rewriting* [81, 82], have brought theorem proving to life for integer multipliers. These methods have solved the automation problem of the previous theorem proving techniques to a large extent.

Term rewriting methods describe the equivalence of semantics of a multiplier design to the multiplication function as a theorem in a theorem prover. Then, they simplify adder and multiplier modules by stating a set of lemmas in the form of equality $lhs = rhs$. These lemmas are used to create a term rewriting mechanism where expressions from circuit definitions are unified with lhs and replaced with their corresponding rhs. If the theorem prover can successfully prove the theorem by a series of term rewriting steps, the correctness of the multiplier is ensured.

Term rewriting methods can verify a wide variety of multipliers when the design hierarchy is available. However, they are not applicable to the gate-level description. Moreover, the library of rewrite rules and lemmas has to be updated for new multiplier architectures. Otherwise, the theorem prover cannot complete the proof.

2.5 Formal Verification Using SCA

In the last six years, SCA-based methods have shown very good results in proving the correctness of large but structurally simple integer multipliers [31, 71, 93, 94]. They have been also employed in verifying floating point multipliers [76], finite field multipliers [37, 66], and dividers [43, 77, 78, 91, 92] as well as debugging faulty integer multipliers [33, 34, 74] and finite field multipliers[36, 67–69].

In this section, we first introduce the basics of SCA. Then, we explain the theory of Gröbner basis and its application in the formal verification of multipliers. Finally, we review the state-of-the-art SCA-based verification methods and highlight their pros and cons.

2.5.1 Definitions

Definition 2.7 A *monomial* is the power product of variables:

$$M = x_1^{\alpha_1} x_2^{\alpha_2} \dots x_n^{\alpha_n} \quad with \quad \alpha_i \in \mathbb{N}_0. \tag{2.3}$$

A monomial with a coefficient is called a *Term*.

Definition 2.8 A *polynomial* is a finite sum of monomials with coefficients in field k:

$$P = c_1 M_1 + c_2 M_2 + \cdots + c_j M_j \quad with \quad c_i \in k. \tag{2.4}$$

The set of all polynomials in $x_1, \ldots, x_n$ with coefficients in k is denoted $k[x_1, \ldots, x_n]$.

The order of monomials in a polynomial can be determined by *lexicographic order*. Assume that the variables are ordered as $x_1 > x_2 > x_3 > \ldots$. Lexicographic order first compares exponents of x_1 in the monomials, in the case of equality it compares exponents of x_2, and so forth. The first monomial (term) after the ordering of P is called *leading monomial* (*leading term*) and denoted by $LM(P)$ $(LT(P))$.

Definition 2.9 A subset $I \subset k[x_1, \ldots, x_n]$ is an *ideal* if it satisfies:

- $0 \in I$.
- If $f, g \in I$, then $f + g \in I$.
- If $f \in I$ and $h \in k[x_1, \ldots, x_n]$, then $hf \in I$.

One of the important examples of an ideal is the ideal generated by a finite number of polynomials.

Definition 2.10 Let $f_1, \ldots, f_s$ be polynomials in $k[x_1, \ldots, x_n]$. Then,

$$\langle f_1, \ldots, f_s \rangle = \left\{ \sum_{i=1}^{s} h_i f_i : h_1, \ldots, h_s \in k[x_1, \ldots, x_n] \right\}, \tag{2.5}$$

where $\langle f_1, \ldots, f_s \rangle$ is the ideal generated by $f_1, \ldots, f_s$. Note that $f_1, \ldots, f_s$ are called a *basis* of the ideal.

As an example, if $f_1 = x - 1 - t$ and $f_2 = y - 1 - t^2$ are the basis of the ideal I, based on Definition 2.10, the following polynomial is an ideal member:

$$x^2 - 2x - y + 2 = (x - 1 + t)(x - 1 - t) + (-1)(y - 1 - t^2) \in I. \tag{2.6}$$

In an *ideal membership* problem, the goal is to decide whether a given polynomial $f \in k[x_1, \ldots, x_n]$ lies in the ideal $\langle f_1, \ldots, f_s \rangle$. For simplicity, we first consider the ideal membership problem for the polynomials with only one variable: Given $f \in k[x]$, to check whether $f \in I = \langle g \rangle$, we divide g into f:

$$f = q \cdot g + r, \tag{2.7}$$

where $q, r \in k[x]$ and $r = 0$ or $deg(r) < deg(g)$. We prove that $f \in I$ if and only if the remainder equals zero, i.e., $r = 0$. Note that the remainder is always unique in this case. For example, suppose we want to check whether the following membership holds:

$$x^3 + 4x^2 + 3x - 7 \in \langle x - 1 \rangle. \tag{2.8}$$

After dividing, we get

$$x^3 + 4x^2 + 3x - 7 = (x^2 + 5x + 8)(x - 1) + 1; \tag{2.9}$$

thus, $x^3 + 4x^2 + 3x - 7$ is not in the ideal $\langle x - 1 \rangle$ since $r \neq 0$.

Now, we consider the ideal membership for multi-variable polynomials: given $f \in k[x_1, \ldots, x_n]$ and an ideal $I = \langle f_1, \ldots, f_s \rangle$, the goal is to determine if $f \in I$. To check the ideal membership, we divide f by $f_1, \ldots, f_s$, i.e., we express f in the form:

$$f = a_1 f_1 + \cdots + a_s f_s + r, \tag{2.10}$$

where the quotients $a_1, \ldots, a_s$ and remainder r lie in $k[x_1, \ldots, x_n]$. If $r = 0$, then we successfully prove that $f \in I$. The division algorithm to obtain the quotients and remainder has been explained in detail in [17]. Unlike the division in $k[x]$ (single variable polynomials), the remainder may change by ordering of the polynomials $(f_1, \ldots, f_s)$ in $k[x_1, \ldots, x_n]$ (multi-variable polynomials). Thus, $r = 0$ is a *sufficient* condition for ideal membership, but it is not a *necessary* condition.

As an example, let $f_1 = xy + 1$, $f_2 = y^2 + 1 \in k[x, y]$. Dividing $f = xy^2 - x$ by $F = (f_1, f_2)$, the result is

$$xy^2 - x = y \cdot (xy + 1) + 0 \cdot (y^2 - 1) + (-x - y); \tag{2.11}$$

however, with a different ordering of polynomials $F = (f_2, f_1)$, we get

$$xy^2 - x = x \cdot (y^2 - 1) + 0 \cdot (xy + 1) + 0. \tag{2.12}$$

The second division in Eq. (2.12) proves the ideal membership, i.e., $f \in \langle f_1, f_2 \rangle$. However, the first division in Eq. (2.11) shows that we might get $r \neq 0$ for an ordering even if the polynomial is a member of the ideal.

Now, an important question arises: is there a "good" generating set for I such that the remainder r on division by "good" generators to be always unique independent of the ordering? In the next section, we see that Gröbner bases have exactly these "good" properties.

2.5.2 *Theory of Gröbner Basis*

Definition 2.11 A finite subset $G = \{g_1, \ldots, g_t\}$ of an ideal I is a *Gröbner basis* if

$$\langle LT(g_1), \ldots, LT(g_t) \rangle = \langle LT(I) \rangle. \tag{2.13}$$

In other words, a set $G = \{g_1, \ldots, g_t\} \subset I$ is a Gröbner basis of I if and only if the leading term of any element of I is divisible by one of the $LT(g_i)$.

The remainder r of dividing f by G is unique no matter how the elements of G are listed. This Gröbner basis property helps us to solve the problem of ideal membership for multi-variable polynomials.

Corollary 2.1 *Let $G = \{g_1, \ldots, g_t\}$ be a Gröbner Basis for an ideal $I \subset k[x_1, \ldots, x_n]$ and let $f \in k[x_1, \ldots, x_n]$. Then, $f \in I$ if and only if the remainder of dividing f by G is zero.*

Before solving an ideal membership problem, it is necessary to make sure whether the generating set is a Gröbner basis. To do so, we first introduce the *S-polynomials* and then present the *Buchberger's Criterion.*

Definition 2.12 The S-polynomial of f and g is as follows:

$$S(f, g) = \frac{LCM(LM(f), LM(g))}{LT(f)} \cdot f - \frac{LCM(LM(f), LM(g))}{LT(g)} \cdot g, \tag{2.14}$$

where LCM is the *least common multiple.*

As an example, the S-polynomial of two polynomials $f = x^3y^2 - x^2y^3$ and $g = 3x^4y + y^2$ with the variable order $x > y > z$ in $\mathbb{R}[x, y]$ is as follows:

$$S(f, g) = \frac{x^4y^2}{x^3y^2} \cdot f - \frac{x^4y^2}{3x^4y} \cdot g = x \cdot f - \frac{1}{3} \cdot y \cdot g = -x^3y^3 - \frac{1}{3}y^3. \tag{2.15}$$

An S-polynomial $S(f, g)$ is designed to generate the cancellation of leading terms. Thus, it can be used in proving if a basis of an ideal is a Gröbner basis.

Theorem 2.1 (Buchnerger's Criterion) *Let I be a polynomial ideal. Then, a basis $G = \{g_1, \ldots, g_t\}$ for I is a Gröbner basis for I if and only if for all pairs $i \neq j$, the remainder of dividing $S(g_i, g_j)$ by G is zero.*

For example, consider the ideal $I = \langle y - x^2, z - x^3 \rangle$ in $\mathbb{R}[x, y, z]$. To prove that $G = \{y - x^2, z - x^3\}$ is a Gröbner basis with the variable order $z > y > x$, we first calculate the S-polynomial

$$S\left(y - x^2, z - x^3\right) = \frac{yz}{y}\left(y - x^2\right) - \frac{yz}{z}\left(z - x^3\right) = -zx^2 + yx^3. \tag{2.16}$$

Then, we find out by dividing that

$$-zx^2 + yx^3 = x^3 \cdot \left(y - x^2\right) + \left(-x^2\right) \cdot \left(z - x^3\right) + 0; \tag{2.17}$$

thus, based on Theorem 2.1, G is a Gröbner basis for I since the remainder is zero.

Algorithm 1 Buchberger's Algorithm

Input: $F = \{f_1, \ldots, f_s\}$
Output: a Gröbner basis $G = (g_1, \ldots, g_t)$ for I, with $F \subset G$
1: $G := F$
2: **repeat**
3: $\quad G' := G$
4: $\quad$ **for each** pair $\{p, q\}$, $p \neq q$ in G' **do**
5: $\quad\quad$ S:=remainder of dividing $S(p, q)$ by G'
6: $\quad\quad$ **if** $S \neq 0$ **then** $G := G \cup \{S\}$
7: **until** $G = G'$
8: **return** G

Theorem 2.1 helps us to check whether a basis is a Gröbner basis. However, we sometimes need to construct a Gröbner basis for the generating set of an ideal.

Theorem 2.2 *Let $I = \langle f_1, \ldots, f_s \rangle \neq \{0\}$ be a polynomial ideal. Then, a Gröbner basis for I can be constructed in a finite number of steps by Algorithm 1.*

Theorem 2.1 helps us to check whether a generating set is a Gröbner basis; if not, then Algorithm 1 is used to construct the Gröbner basis from the initial generating set. Theorem 2.1 and Algorithm 1 can be used in the formal verification of multipliers for detecting and constructing Gröbner bases. However, they usually require many time-consuming steps, including several computations of S-polynomials. Thus, we take advantage of Theorem 2.3, which significantly simplifies and speeds up the process.

Theorem 2.3 *Given a finite set $G \subset k[x_1, \ldots, x_n]$, suppose that we have $f, g \in G$ such that*

$$LCM(LM(f), LM(g)) = LM(f) \cdot LM(g). \tag{2.18}$$

In other words, the leading monomials of f and g are relatively prime. Then, $S(f, g) \xrightarrow{G} 0$, i.e., the remainder of dividing $S(f, g)$ by G is zero.

It can be instantly concluded from Theorem 2.3 that if all members of a basis are relatively prime, the basis is a Gröbner basis. For example, let $G = \{x^2 + y, z^4\}$ with the variable order $x > y > z$ on $\mathbb{R}[x, y, z]$. The leading monomials of G members, i.e., x^2 and z^4, are relatively prime. Thus, G is a Gröbner basis.

The proofs of Theorem 2.1, Theorem 2.2, and Theorem 2.3 can be found in [2, 17]. In the next section, we explain how the ideal membership and Gröbner basis definitions are used in the SCA-based formal verification of multipliers.

2.5.3 SCA-Based Verification

In SCA-based verification, the gate-level netlist (or AIG) and the *Specification Polynomial* (SP) are given as inputs, and the task is to formally prove that the SP and the multiplier are equivalent. The SP is a polynomial determining the word-level function of a multiplier based on its inputs and outputs. For an $N \times N$ unsigned integer multiplier with $A_{N-1}A_{N-2}\ldots A_0$ and $B_{N-1}B_{N-2}\ldots B_0$ inputs and $Z_{2N-1}Z_{2N-2}\ldots Z_0$ output, the SP is

$$SP = \sum_{i=0}^{2N-1} 2^i Z_i - \left(\sum_{i=0}^{N-1} 2^i A_i\right) \times \left(\sum_{i=0}^{N-1} 2^i B_i\right). \tag{2.19}$$

For signed multipliers using two's complement, the SP is equal to

$$SP = -2^{2N-1} Z_{2N-1} + \sum_{i=0}^{2N-2} 2^i Z_i$$
$$-\left(-2^{N-1} A_{N-1} + \sum_{i=0}^{N-2} 2^i A_i\right) \times \left(-2^{N-1} B_{N-1} + \sum_{i=0}^{N-2} 2^i B_i\right). \tag{2.20}$$

For example, the SP for the 2×2 unsigned multiplier of Fig. 2.1 is $SP = 8Z_3 + 4Z_2 + 2Z_1 + Z_0 - (2A_1 + A_0) \times (2B_1 + B_0)$, where $8Z_3 + 4Z_2 + 2Z_1 + Z_0$ shows the world-level representation of the 4-bit output, and $(2A_1 + A_0) \times (2B_1 + B_0)$ indicates the product of the 2-bit inputs. One the other hand, The SP for a 2×2 signed multiplier is $SP = -8Z_3 + 4Z_2 + 2Z_1 + Z_0 - (-2A_1 + A_0) \times (-2B_1 + B_0)$.

The gates of a circuit (or nodes of an AIG) can be captured as polynomials describing the word-level relation between their inputs and output. In the gate-level representation of a circuit, each logic gate with output z and inputs A and B performs one of the eight basic operations in Table 2.1. The polynomials describing the function of each gate are shown in Table 2.3. The variables in these polynomials are always ordered based on the reverse topological order of the circuit, i.e., a gate's output is always in a higher order than its inputs.

The gate polynomials for the gate-level representation of the 2×2 multiplier in Fig. 2.1a are as follows:

$$\begin{aligned}
P_{g_8} &= Z_3 - w_1 w_4, & P_{g_7} &= Z_2 - w_1 - w_4 + 2w_1 w_4,\\
P_{g_6} &= w_4 - w_2 w_3, & P_{g_5} &= Z_1 - w_2 - w_3 + 2w_2 w_3,\\
P_{g_4} &= w_1 - A_1 B_1, & P_{g_3} &= w_2 - A_1 B_0,\\
P_{g_2} &= w_3 - A_0 B_1, & P_{g_1} &= w_2 - A_0 B_0,
\end{aligned} \tag{2.21}$$

Table 2.3 Logic gates polynomials

Type	Boolean algebra	Polynomial
BUF	$z = A$	$P_G = z - A$
NOT	$z = \overline{A}$	$P_G = z - 1 + A$
AND	$z = A \wedge B$	$P_G = z - AB$
OR	$z = A \vee B$	$P_G = z - A - B + AB$
NAND	$z = \overline{A \wedge B}$	$P_G = z - 1 + AB$
NOR	$z = \overline{A \vee B}$	$P_G = z - 1 + A + B - AB$
XOR	$z = A \oplus B$	$P_G = z - A - B + 2AB$
XNOR	$z = A \odot B$	$P_G = z - 1 + A + B - 2AB$

Table 2.4 AIG nodes polynomials

Boolean algebra	Node	Polynomial
$z = n_i$	$n_i \longrightarrow z$	$P_N = z - n_i$
$z = \overline{n_i}$	$n_i \dashrightarrow z$	$P_N = z - 1 + n_i$
$z = n_i \wedge n_j$	n_i, z, n_j	$P_N = z - n_i n_j$
$z = \overline{n_i} \wedge n_j$	n_i, z, n_j	$P_N = z - n_j + n_i n_j$
$z = \overline{n_i} \wedge \overline{n_j}$	n_i, z, n_j	$P_N = z - 1 + n_i + n_j - n_i n_j$

where the polynomials for the AND gates (g_1, g_2, g_3, g_4, g_6, and g_8) and XOR gates (g_5 and g_7) are captured based on Table 2.3.

In the AIG representation of a circuit, there are only five possible Boolean operations in each node with respect to the input edge negations. The polynomials for these operations are presented in Table 2.4, where z is the output of a node and n_i and n_j are the inputs. Note that the first two operations in Table 2.4 only happen on the primary output of a multiplier, where the node outputs are connected to the primary outputs with or without negation. Moreover, the variables in these polynomials are ordered based on the reverse topological order of the circuit, i.e., a node's output is always in a higher order than its inputs.

The captured polynomials for the AIG representation of the 2×2 multiplier in Fig. 2.1b are as follows:

$$\begin{aligned}
P_{Z_3} &= Z_3 - n_{11}, & &\dots \\
P_{Z_2} &= Z_2 - 1 + n_{12}, & P_{n_3} &= n_3 - A_1B_0, \\
P_{n_{12}} &= n_{12} - 1 + n_9 + n_{10} - n_9n_{10}, & P_{n_2} &= n_2 - A_0B_1, \\
&\dots & P_{n_1} &= n_1 - A_0B_0.
\end{aligned} \tag{2.22}$$

With specification polynomial and gates/nodes polynomials in hand, we can verify the multiplier using ideal membership.

Theorem 2.4 *Let $I = \langle P_1, \dots, P_s \rangle$, where $P_1, \dots, P_s$ are the gates/nodes polynomials. The multiplier is correct if and only if the specification polynomial SP is a member of I, i.e., $SP \in I$.*

Using reverse topological order of the circuit, the leading monomials of all $G = \{P_1, \dots, P_s\}$ members are relatively prime, e.g., in Eqs. (2.21) and (2.22), the leading monomials always consist of only one variable which is the output of a gate/node, and since the gates/nodes outputs are unique, the leading monomials are relatively prime. Therefore, based on Theorem 2.3, G is a Gröbner basis for I. Subsequently, the ideal membership of SP in I is proven by division. If the remainder of dividing SP by $P_1, \dots, P_s$ is zero, the circuit is correct; otherwise, it is buggy. The proof of Theorem 2.4 and the details of SCA-based verification can be found in [42, 71].

For example, the correctness of the 2×2 gate-level multiplier of Fig. 2.1a is proven by the step-wise division of SP by the gate polynomials (see Eq. (2.21)) as follows:

$$\begin{aligned}
&SP := 8Z_3 + 4Z_2 + 2Z_1 + Z_0 - (4A_1B_1 + 2A_1B_0 + 2A_0B_1 + A_0B_0), \\
&SP \xrightarrow{p_{g8}} SP_1 := 8w_1w_4 + 4Z_2 + 2Z_1 + Z_0 - (4A_1B_1 + 2A_1B_0 + 2A_0B_1 + A_0B_0), \\
&SP_1 \xrightarrow{p_{g7}} SP_2 := 4w_1 + 4w_4 + 2Z_1 + Z_0 - (4A_1B_1 + 2A_1B_0 + 2A_0B_1 + A_0B_0), \\
&SP_2 \xrightarrow{p_{g6}} SP_3 := 4w_1 + 4w_2w_3 + 2Z_1 + Z_0 - (4A_1B_1 + 2A_1B_0 + 2A_0B_1 + A_0B_0), \\
&SP_3 \xrightarrow{p_{g5}} SP_4 := 4w_1 + 2w_2 + 2w_3 + Z_0 - (4A_1B_1 + 2A_1B_0 + 2A_0B_1 + A_0B_0), \\
&SP_4 \xrightarrow{p_{g4}} SP_5 := 2w_2 + 2w_3 + Z_0 - (2A_1B_0 + 2A_0B_1 + A_0B_0), \\
&SP_5 \xrightarrow{p_{g3}} SP_6 := 2w_3 + Z_0 - (2A_0B_1 + A_0B_0), \\
&SP_6 \xrightarrow{p_{g2}} SP_7 := Z_0 - (A_0B_0), \\
&SP_7 \xrightarrow{p_{g1}} r := 0,
\end{aligned} \tag{2.23}$$

where the remainder r equals zero; thus, the multiplier is bug-free.

As another example, the AIG representation of the 2×2 multiplier of Fig. 2.1b is verified by the step-wise division of SP by the node polynomials (see Eq. (2.22)) as follows:

$$
\begin{aligned}
&SP := 8Z_3 + 4Z_2 + 2Z_1 + Z_0 - (4A_1B_1 + 2A_1B_0 + 2A_0B_1 + A_0B_0),\\
&SP \xrightarrow{P_{Z_3}} SP_1 := 8n_{11} + 4Z_2 + 2Z_1 + Z_0 - (4A_1B_1 + 2A_1B_0 + 2A_0B_1 + A_0B_0),\\
&SP_1 \xrightarrow{P_{Z_2}} SP_2 := 8n_{11} + 4 - 4n_{12} + 2Z_1 + Z_0 - (4A_1B_1 + 2A_1B_0 + 2A_0B_1 + A_0B_0),\\
&\quad\vdots\\
&SP_{13} \xrightarrow{P_{n_3}} SP_{14} := n_2 + n_1 - (A_0B_1 + A_0B_0),\\
&SP_{14} \xrightarrow{P_{n_2}} SP_{15} := n_1 - (A_0B_0),\\
&SP_{15} \xrightarrow{P_{n_1}} r := 0,
\end{aligned}
\tag{2.24}
$$

where the remainder r is equal to zero; thus, the multiplier is correct.

In the verification of integer multipliers, all variables in polynomials are Boolean. Thus, we always replace powers $x_i^{a_i}$ with $a_i > 1$ by x_i, since x_i can only take values from $\{0, 1\}$. Furthermore, dividing SP_i by a gate/node polynomial $P_N = x - tail(P_N)$ is equivalent to *substituting* x with $tail(P_N)$ in SP_i. For example, to obtain the result of the first division step in Eq. (2.24), Z_3 can be substituted with n_{11} in SP to obtain SP_1. The process of dividing the SP by gate/node polynomials (or equivalently, substituting gate/node polynomials in SP) is called *backward rewriting*. We always prefer substitution over division as the substitution is less expensive in terms of run-time [14]. We refer to the intermediate polynomials during backward rewriting as SP_i in the rest of this book.

2.5.4 State-of-the-art of SCA-Based Verification Methods

In the last six years, several SCA-based methods have been introduced to verify integer multipliers. These methods have improved the performance (i.e., speed up the process) or the applicability (i.e., support a wider variety of architectures) of the basic SCA-based verification introduced in this chapter. The authors of [15, 93] proposed a method to capture the gate-level netlist as a set of polynomials; then, substitute these polynomials in the specification polynomial step-by-step following the reverse topological order of the circuit.

The work of [31, 32] divides the netlist into fanout-free cones and extracts the polynomial for each cone. Subsequently, it uses the cone polynomials instead of the gate polynomials in substitution steps to reduce the total number of generated monomials during backward rewriting. This approach speeds up the verification of structurally simple multipliers.

The column-wise method of [42, 71] cuts the circuit into slices and verifies the correctness incrementally. The just mentioned approaches have two main disadvantages: (1) They extract the polynomials for the smallest building blocks of a multiplier, i.e., gates; thus, these methods are unaware of building blocks, e.g., half-adders and full-adders, having more compact polynomials. (2) They do not support the verification of structurally complex multipliers, including optimized architectures, since an explosion happens in the size of intermediate polynomials during backward rewriting.

The proposed approaches in [72, 94] take advantage of reverse engineering to identify half-adders and full-adders in the AIG representation of a multiplier. Then, they use the compact polynomials of half-adders and full-adders during backward rewriting, which speeds up the verification process significantly. However, these approaches still do not provide any solution to avoid explosion during the backward rewriting of complex multipliers, which confines their applicability to structurally simple designs. Furthermore, they do not support the detection of larger building blocks such as compressors.

The work of [75] aims to attack the explosion problem during the verification of structurally complex multipliers. It presents an XOR rewriting technique that groups the gates into cones based on the XOR gates. Then, it extracts the polynomials for each cone and removes the redundant monomials known as vanishing monomials. The method works for some complex architectures. However, it is not robust since it misses many vanishing monomials. Moreover, the approach does not investigate the origin of vanishing monomials in structurally complex multipliers.

The proposed method in [41] uses a combination of SAT and SCA to verify structurally complex multipliers. The authors come up with an algorithm to detect the FSA in a multiplier and verify it using SAT. Then, the adder is substituted with an architecturally simple adder. Finally, the SCA-based verification is performed on the new architecture. The method achieves very good run-times if the FSA can be detected, which is not always possible, particularly for optimized multipliers.

Chapter 3
Challenges of SCA-Based Verification

This chapter investigates the results of applying SCA-based verification to different multiplier architectures. It has been experimentally shown that SCA-based verification reports very good results for a small set of simple multipliers. However, it fails when it comes to verifying complex architectures. In this chapter, the reasons for this failure and the solutions to overcome them are studied thoroughly.

This chapter first introduces structurally simple multipliers and provides the results of applying SCA-based verification to prove their correctness. Subsequently, structurally complex multipliers are introduced, and the challenges of verifying them are discussed. Finally, the three techniques, i.e., local vanishing monomials removal, reverse engineering, and dynamic backward rewiring, are briefly introduced to overcome the challenges.

3.1 Introduction

While the BDD-based verification methods run quickly out of memory and the SAT-based verification techniques run out of time even for small multipliers, SCA-based verification can successfully prove the correctness of some large multiplier architectures. Over the last few years, several techniques have been proposed to speed up the backward rewriting process. They have significantly improved the scalability of SCA-based verification and successfully verified huge multipliers with millions of gates, e.g., the proposed technique in [94] verifies a 512×512 multiplier in approximately 30 s.

Despite these improvements, the SCA-based methods can only verify a very limited set of multiplier architectures. Therefore, they might prove the correctness of a very big 512×512 multiplier in a few seconds, but they fail to verify another small architecture that implements the 12×12 multiplication. Now, some questions arise regarding SCA-based verification:

A. Mahzoon et al., *Formal Verification of Structurally Complex Multipliers*,
https://doi.org/10.1007/978-3-031-24571-8_3

- Which architectures can be verified using SCA-based verification? and which architectures cannot be verified? What are the properties of these architectures?
- How can SCA-based verification be extended in order to verify a wider variety of multiplier architectures?

To answer these questions, we group multipliers into two categories: structurally simple and structurally complex multipliers. In structurally simple multipliers, the second and third stages are only made of half-adders and full-adders. On the other hand, in structurally complex multipliers, there are several extra logic gates (or nodes) in addition to half-adders and full-adders. These multipliers are either generated by sophisticated algorithms, which result in highly parallel architectures, particularly for the FSA, or obtained by optimizing structurally simple multipliers.

The basic SCA-based verification method introduced in Sect. 2.5.3 and its improvements [31, 71, 72, 93, 94] can only prove the correctness of structurally simple multipliers. However, when it comes to the verification of structurally complex multipliers, they fail even for small architectures. This problem can be effectively solved by integrating three techniques into the basic SCA-based verification: local vanishing monomials removal, reverse engineering, and dynamic backward rewiring.

This chapter provides a detailed description of structurally simple and structurally complex multipliers. Moreover, it provides several experimental results to illustrate the details of SCA-based verification when it comes to verifying different architectures. We also provide a brief explanation of our three techniques in order to solve the problem of verifying structurally complex multipliers. These techniques are explained in detail in the next three chapters.

3.2 Verification of Structurally Simple Multipliers

In this section, we first provide the definition of structurally simple multipliers. Then, we present the results of verifying these architectures using SCA-based verification. Finally, we discuss the experimental results and give insight into the success of SCA-based verification in proving the correctness of structurally simple multipliers.

3.2.1 Definition of Structurally Simple Multipliers

Definition 3.1 A multiplier is *structurally simple* if its second and third stages are only made of half-adders and full-adders.

In the second stage of a multiplier, partial products are reduced using multi-operand adders. Most of the proposed architectures for this stage use only half-adder and full-adders as multi-operand adders, e.g., array, Wallace tree, and Dadda tree.

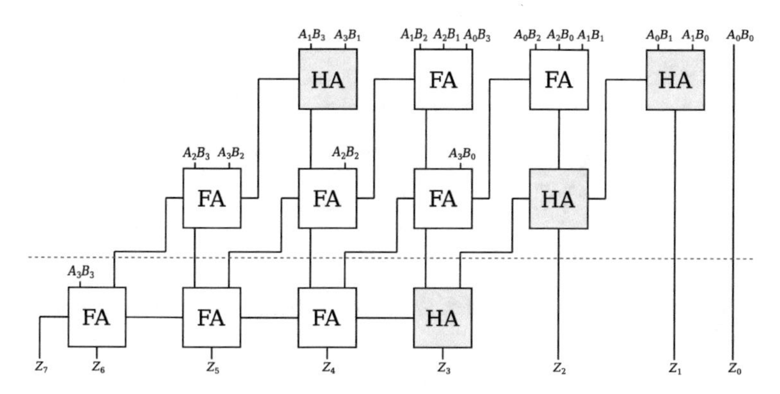

Fig. 3.1 4×4 multiplier with $SP \circ AR \circ RC$ architecture

The compressor tree is one of the few PPA architectures that takes advantage of (4:2) compressors and thus cannot be fully implemented using half-adders and full-adders.

On the other hand, the ripple carry adder is the only architecture for the third stage of a multiplier which is fully made of half-adders and full-adders. The other FSA architectures have some extra gates (or nodes) in addition to half-adder and full-adders.

Figure 3.1 shows a 4×4 multiplier with $SP \circ AR \circ RC$ architecture (SP: simple PPG, AR: array, and RC: ripple carry adder). The first stage has been omitted for the sake of simplicity; thus, $A_i B_j$ depicts a partial product. The dashed lines separate the second and third stages of the multiplier. These stages are only made of half-adders and full-adders. Thus, the multiplier is structurally simple.

3.2.2 Experimental Results

In this section, we investigate the efficiency of basic SCA-based verification in proving the correctness of structurally simple multipliers. To achieve this goal, we provide experimental evidence for the verification of three types of structurally simple multipliers. These multiplier architectures are:

1. *Simple partial product generator* $\circ$ *array* $\circ$ *ripple carry adder* ($SP \circ AR \circ RC$)
2. *Simple partial product generator* $\circ$ *Wallace tree* $\circ$ *ripple carry adder* ($SP \circ WT \circ RC$)
3. *Simple partial product generator* $\circ$ *Dadda tree* $\circ$ *ripple carry adder* ($SP \circ DT \circ RC$)

For a comprehensive evaluation, the results of SCA-based verification for these multipliers are reported in two different types of graphs:

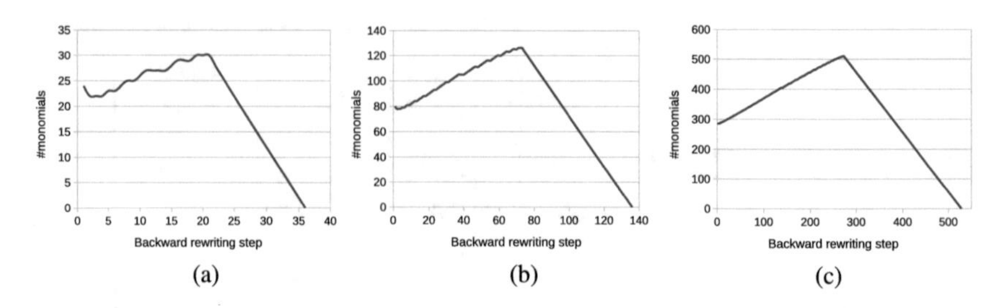

Fig. 3.2 Size of intermediate polynomials SP_i during the backward rewriting of $SP \circ AR \circ RC$. (**a**) 4×4. (**b**) 8×8. (**c**) 16×16

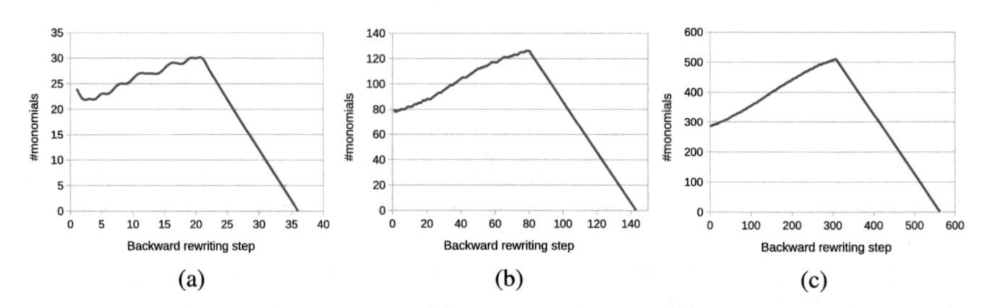

Fig. 3.3 Size of intermediate polynomials SP_i during backward the rewriting of $SP \circ WT \circ RC$. (**a**) 4×4. (**b**) 8×8. (**c**) 16×16

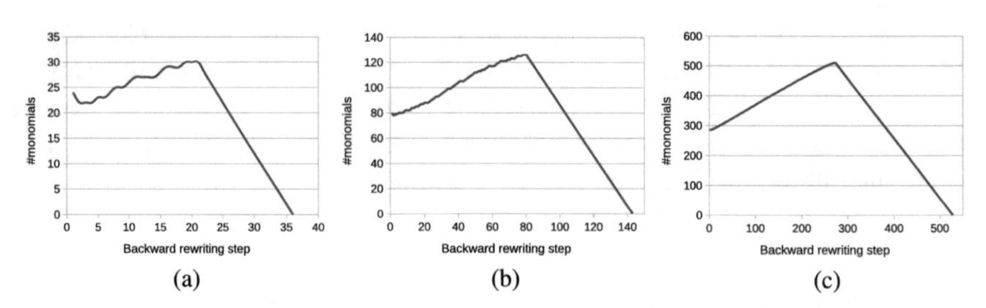

Fig. 3.4 Size of intermediate polynomials SP_i during the backward rewriting of $SP \circ DT \circ RC$. (**a**) 4×4. (**b**) 8×8. (**c**) 16×16

- The size of intermediate polynomials SP_i in each step of backward rewriting is reported for 4×4, 8×8, and 16×16 multipliers. Figures 3.2, 3.3, and 3.4 present the graphs for $SP \circ AR \circ RC$, $SP \circ WT \circ RC$, and $SP \circ DT \circ RC$ architectures, respectively.
- The maximum size of intermediate polynomials SP_i during backward rewriting is reported for the different multiplier sizes. Figure 3.5 shows the graphs for $SP \circ AR \circ RC$, $SP \circ WT \circ RC$, and $SP \circ DT \circ RC$ architectures, respectively.

In the first experiment (see Figs. 3.2, 3.3, and 3.4), the size of intermediate polynomial for all structurally simple multipliers grows slightly (i.e., always less than $2\times$ of the initial SP size). Then, it starts to decrease in the middle steps until it eventually becomes one. Note that all multipliers considered here are correct; hence,

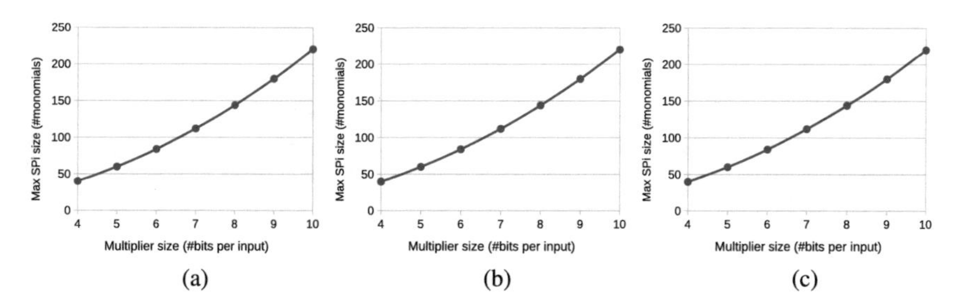

Fig. 3.5 Maximum size of intermediate polynomials SP_i for different structurally simple multiplier sizes. (**a**) $SP \circ AR \circ RC$. (**b**) $SP \circ WT \circ RC$. (**c**) $SP \circ DT \circ RC$

the final result is the zero polynomial containing only one monomial which is 0. For example, for the 16×16 multiplier with $SP \circ AR \circ RC$ architecture, the number of monomials grows slightly until it reaches $1.77\times$ of the initial SP size. Then, it starts to decrease at the 274th step until it eventually becomes one at the 528th step. It is evident from the first experiment that no explosion (i.e., sudden increase in the size of intermediate polynomials) happens during the verification of structurally simple multipliers.

In the second experiment (see Fig. 3.5), the maximum size of intermediate polynomials grows polynomially (almost quadratically) with respect to the multiplier size for the three structurally simple multipliers. Thus, it confirms that SCA-based verification is scalable when it comes to verifying structurally simple multipliers.

3.2.3 *Discussion*

In this section, we give insight into the success of SCA-based verification when it is applied to structurally simple multipliers. We clarify the role of half-adders and full-adders during the backward rewriting of these multipliers.

Half-adders and full-adders have a compact word-level relation between their inputs and outputs:

$$\begin{aligned} HA(\text{in: } X, Y \quad \text{out: } C, S) \quad &\Rightarrow \quad 2C + S = X + Y, \\ FA(\text{in: } X, Y, Z \quad \text{out: } C, S) \quad &\Rightarrow \quad 2C + S = X + Y + Z. \end{aligned} \tag{3.1}$$

If SP_i contains the output polynomial (i.e., $2kC + kS$) and no other occurrences of C and S, the output polynomial can be substituted with the input polynomial (i.e., $kX + kY$ for the half-adder and $kX + kY + kZ$ for the full-adder). Substituting the output polynomial of a half-adder (full-adder) increases the size of SP_i (i.e., the number of monomials) by zero (one), respectively.

Since the third and second stages of a structurally simple multiplier are only made of half-adders and full-adders, each step of backward rewriting increases the size of

SP_i by at most one monomial. Then, the substitution of AND gates' polynomials in the first stage of the multiplier reduces the size of SP_i by two monomials due to term cancellations. It is in line with the experimental results in Figs. 3.2, 3.3, and 3.4. In these experiments, the number of monomials increases slightly during the backward rewriting of the third and second stages of the multipliers. Subsequently, it starts to decrease during the backward rewriting of the first stage, until the zero monomial, i.e., remainder, is obtained.

In this section, we always assumed that the boundaries of half-adders and full-adders are available. Thus, their polynomials are immediately substituted in SP_i when they are visited during backward rewriting. It is also possible to substitute the gates/nodes polynomials of half-adders and full-adders. In this case, the number of substitution steps increases, and some local growth happen during the substitution of gates/nodes polynomials of a half-adder or a full-adder. However, SCA-based verification is still carried out successfully for structurally simple multipliers.

3.3 Verification of Structurally Complex Multipliers

In this section, we provide the definition of structurally complex multipliers. Then, we give the results of verifying these multipliers using SCA-based verification. Finally, we discuss the experimental results.

3.3.1 Definition of Structurally Complex Multipliers

Definition 3.2 A multiplier is *structurally complex* if its second and third stages are not only made of half-adders and full-adders.

Structurally complex multipliers are the results of (1) employing a sophisticated algorithm to implement the second or third stage of a multiplier or (2) applying an optimization algorithm to reduce area or delay.

3.3.1.1 Sophisticated Multiplication Algorithms

Several architectures have been proposed, particularly for the third stage of a multiplier in order to speed up the multiplication process. In contrast to the ripple carry adder, they take advantage of extra hardware to compute the carry bits as fast as possible. Therefore, they consist of several extra gates/nodes in addition to half-adders and full-adders.

Figure 3.6 shows a 4×4 multiplier with $SP \circ AR \circ CL$ architecture (SP: simple PPG, AR: array, and CL: carry look-ahead adder). The first stage has been omitted for the sake of simplicity; thus, $A_i B_j$ depicts a partial product. The dashed lines

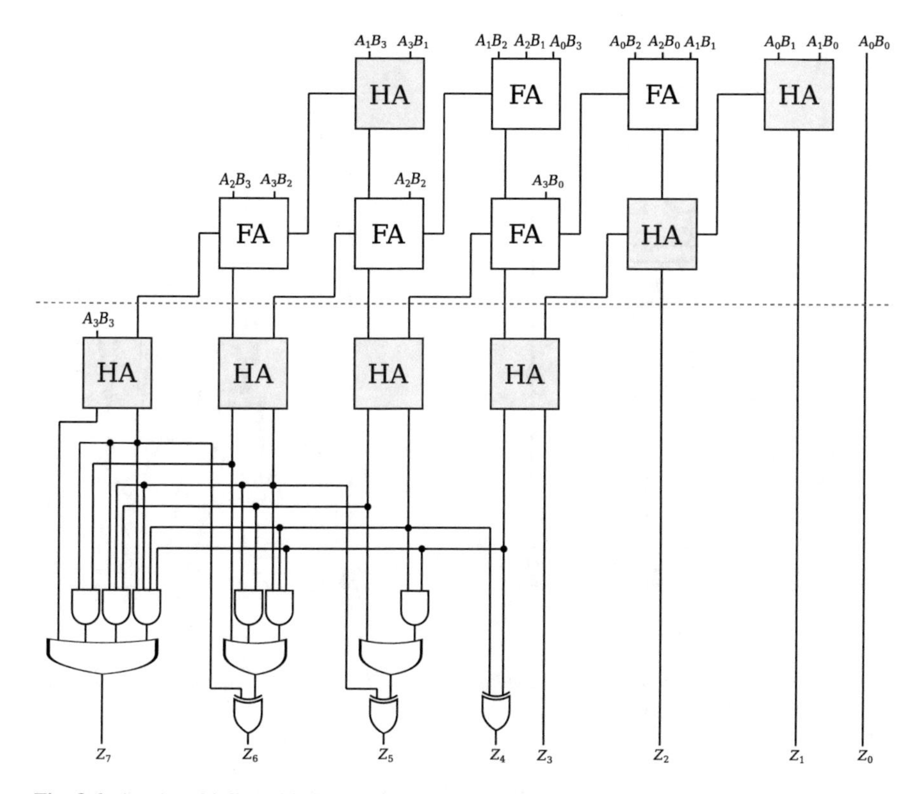

Fig. 3.6 4 × 4 multiplier with $SP \circ AR \circ CL$ architecture

separate the second and third stages of the multiplier. The third stage has several extra AND, OR, and XOR gates in addition to the four half-adders. Hence, the multiplier is structurally complex.

3.3.1.2 Optimization

Optimization of digital circuits is one of the trending research areas in recent years. Several optimization algorithms have been proposed to reduce the size of digital circuits or increase their speed [56–58]. Most of these algorithms are applicable to the AIG representation since some AIG concepts (e.g., cuts) facilitate the optimization process.

The effects of optimization on the network of half-adders and full-adders in the second and third stages of a multiplier appear at two hierarchy levels:

1. Logic optimization inside half-adders and full-adders: The number of nodes in half-adders and full-adders shrinks, e.g., 11 AIG nodes of a full-adder are reduced to 8 nodes. In this case, the overall structure of the multiplier does not

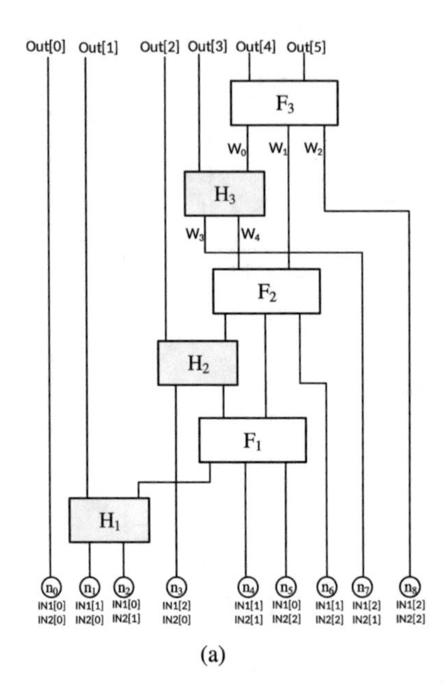

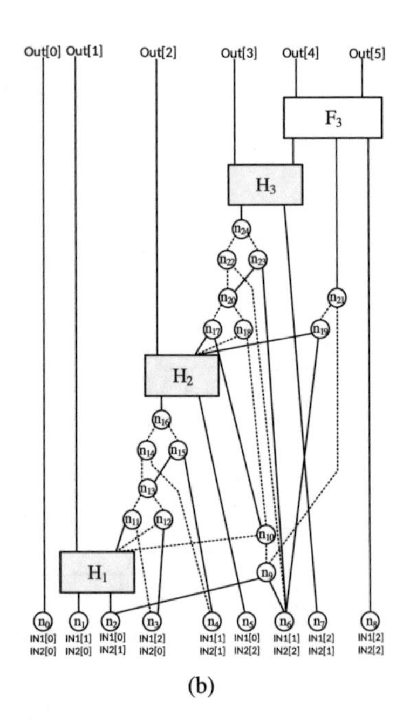

Fig. 3.7 AIG representations of 3 × 3 multipliers with $SP \circ AR \circ RC$ architecture. (**a**) Before optimization. (**b**) After optimization

change significantly; thus, if the multiplier is structurally simple, it still remains structurally simple after the optimization.

2. Logic optimization across half-adders and full-adders: Half-adders and full-adders are merged to reduce the number of nodes, and as a consequence, some boundaries are removed. In this case, the second and third stages of the multiplier no longer consist of only half-adders and full-adders. Thus, if the multiplier is structurally simple, it becomes structurally complex after the optimization.

Figure 3.7a shows the AIG of a 3 × 3 multiplier with $SP \circ AR \circ RC$ architecture. The first stage consists of nine AND gates (i.e., $n_0, n_1, \ldots, n_8$) to generate partial products. The second and third stages are made of half-adder and full-adder networks to reduce partial products and perform the final addition. As can be seen in Fig. 3.7a, the boundaries of half-adders and full-adders and the connections between them are fully visible.

Figure 3.7b shows the AIG of the 3 × 3 multiplier after applying *resyn3* optimization from abc [1]. While the overall number of AIG nodes is reduced by 15%, the optimization destroys the boundaries of the two full-adders F_1 and F_2. In addition to half-adders and full-adders, there are now several extra nodes in the second and third stages of the multiplier; thus, it is structurally complex.

3.3.2 Experimental Results

In this section, we study the efficiency of basic SCA-based verification in proving the correctness of structurally complex multipliers. To do this, we provide experimental evidence for the verification of three types of structurally complex multipliers. These multiplier architectures are:

1. *Simple partial product generator* $\circ$ *array* $\circ$ *carry look-ahead adder* ($SP \circ AR \circ CL$)
2. *Simple partial product generator* $\circ$ *Wallace tree* $\circ$ *Ladner–Fischer adder* ($SP \circ WT \circ LF$)
3. *Simple partial product generator* $\circ$ *Dadda tree* $\circ$ *Brent-Kung adder* ($SP \circ DT \circ BK$)

For a detailed evaluation, the results of SCA-based verification for these multipliers are reported in two different types of graphs:

- The size of intermediate polynomials SP_i in each step of backward rewriting is reported for 4×4, 8×8, and 16×16 multipliers. Figures 3.8, 3.9, and 3.10 present the graphs for $SP \circ AR \circ CL$, $SP \circ WT \circ LF$, and $SP \circ DT \circ BK$ architectures, respectively.
- The maximum size of intermediate polynomial SP_i during backward rewriting is reported for the different multiplier sizes. Figure 3.11 shows the graphs for $SP \circ AR \circ CL$, $SP \circ WT \circ LF$, and $SP \circ DT \circ BK$ architectures, respectively.

In the first experiment (see Figs. 3.8, 3.9, and 3.10), the size of intermediate polynomials SP_i grows dramatically several times during the backward rewriting of structurally complex multipliers:

- For the $SP \circ AR \circ CL$ multipliers in Fig. 3.8 with 4×4 and 8×8 sizes, the maximum number of monomials reaches $2.0\times$ and $72.5\times$ compared to the initial number of monomials, respectively. The situation is even worse for the 16×16 multiplier since the number of monomials explodes after about 140 steps of substitution.
- For the $SP \circ WT \circ LF$ multipliers in Fig. 3.9 with 4×4 and 8×8 sizes, the maximum number of monomials reaches $1.3\times$ and $13.1\times$ compared to the initial number of monomials, respectively. The verification of the 16×16 multiplier fails since an explosion happens in the number of monomials after about 180 steps of substitution.
- For the $SP \circ DT \circ BK$ multipliers in Fig. 3.10 with 4×4 and 8×8 sizes, the maximum number of monomials reaches $1.7\times$ and $30{,}017.1\times$ compared to the initial number of monomials, respectively. Similar to the two previous multipliers, the verification of the 16×16 multiplier fails due to an explosion in the number of monomials after about 120 steps of substitution.

In the second experiment (see Fig. 3.11), the maximum size of intermediate polynomials grows exponentially with respect to the multiplier size for the three

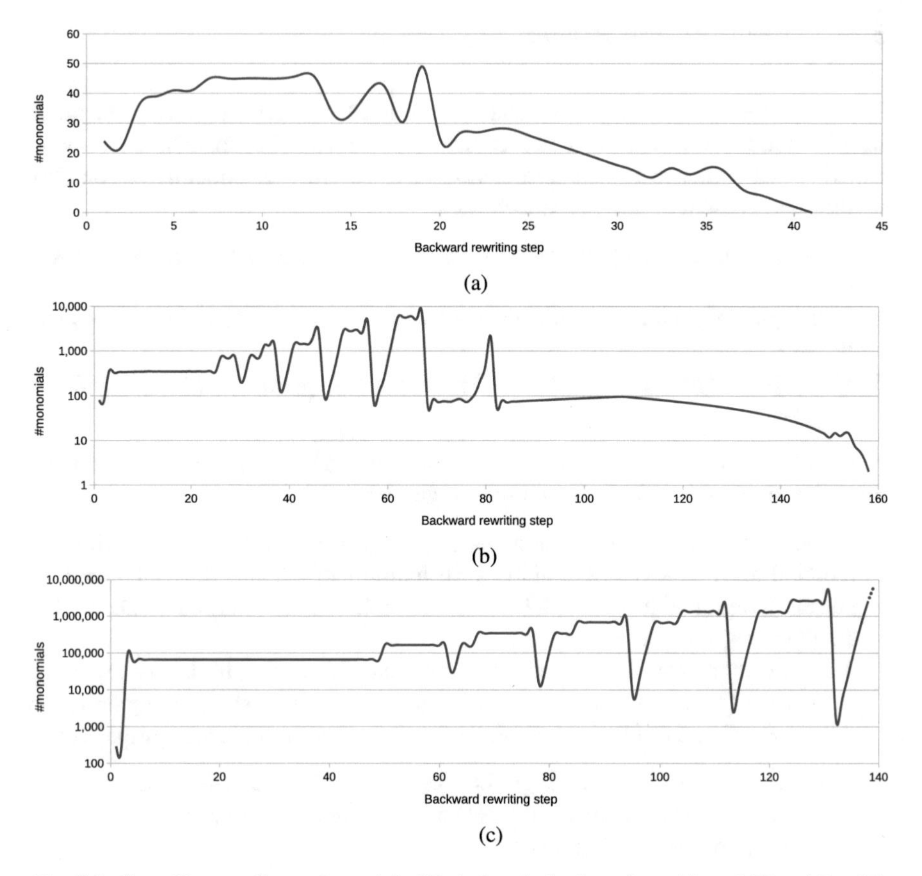

Fig. 3.8 Size of intermediate polynomials SP_i during the backward rewriting of $SP \circ AR \circ CL$. (**a**) 4×4. (**b**) 8×8. (**c**) 16×16

structurally complex multipliers. Therefore, it can be concluded that SCA-based verification is not scalable when it comes to the verification of structurally complex multipliers.

3.3.3 *Discussion*

In this section, we give more insight into the monomials explosion during the backward rewriting of structurally complex multipliers.

As we showed in Sect. 3.2.3, the substitution of the half-adders and full-adders polynomials increases the size of intermediate polynomials SP_i by at most one monomial. However, in structurally complex multipliers, there are several extra gates/nodes in the second and third stages of the multiplier in addition to half-adders and full-adders. Sudden increases in the size of intermediate polynomials

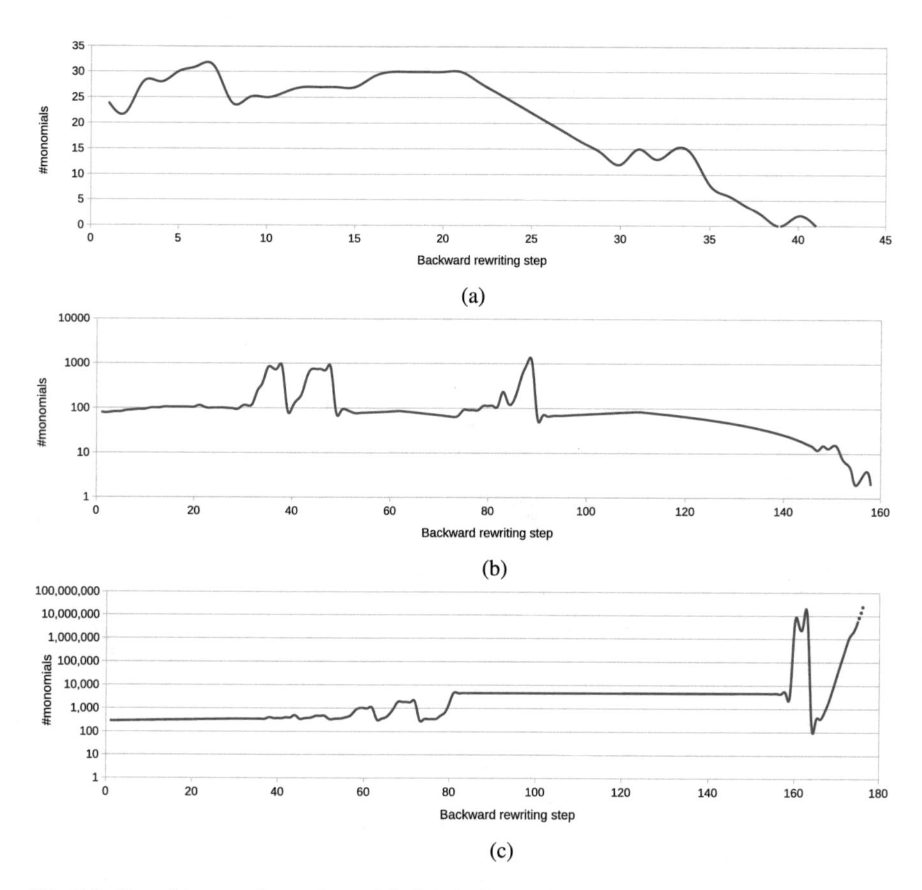

Fig. 3.9 Size of intermediate polynomials SP_i during the backward rewriting of $SP \circ WT \circ LF$. (**a**) 4×4. (**b**) 8×8. (**c**) 16×16

usually happen when these extra gates/nodes are met during backward rewriting. More precisely, the extra gates/nodes cause two major problems, which are the source of monomials explosion during backward rewriting:

- Generation of redundant monomials known as *vanishing monomials*. These monomials are generated during the verification and reduced to zero after several substitution steps. However, the huge number of vanishing monomials before cancellation causes a blow-up in computations.
- Ordering of gates/nodes during backward rewriting. There are usually many possible orderings for the substitution of the extra gates/nodes polynomials. Some of these orderings increase the size of SP_i dramatically and even result in an explosion.

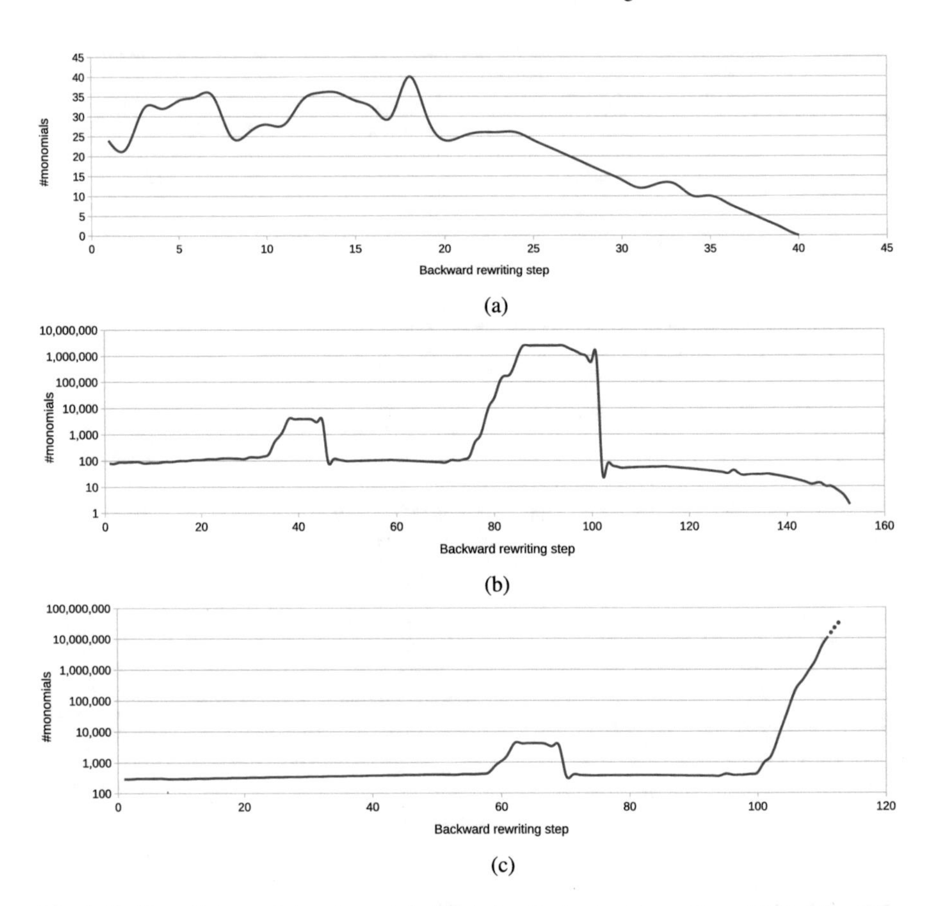

Fig. 3.10 Size of intermediate polynomials SP_i during the backward rewriting of $SP \circ DT \circ BK$. (**a**) 4×4. (**b**) 8×8. (**c**) 16×16

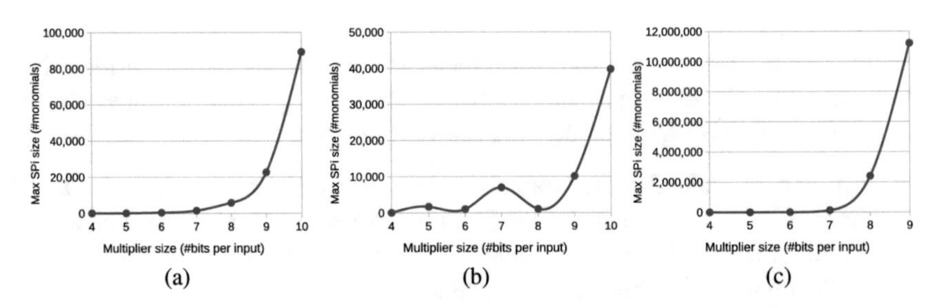

Fig. 3.11 Maximum size of intermediate polynomials SP_i for different structurally complex multiplier sizes. (**a**) $SP \circ AR \circ CL$. (**b**) $SP \circ WT \circ LF$. (**c**) $SP \circ DT \circ BK$

In the next three chapters, we investigate the aforementioned problems in depth and come up with some solutions to overcome them. An overview of these solutions is presented in the following section.

3.4 Overcoming the Challenges

We now briefly introduce our three techniques to extend the basic SCA-based verification and overcome the challenges of verifying structurally complex multipliers. Our proposed techniques are:

- **Local vanishing monomials removal:** detecting and removing vanishing monomials before global backward rewriting
- **Reverse engineering:** identifying atomic blocks, i.e., half-adders, full-adders, and compressors, to facilitate the detection of vanishing monomials and speed up global backward rewriting
- **Dynamic backward rewriting:** using a dynamic substitution order during backward rewriting, which allows controlling the size of intermediate polynomials

In the next three chapters, we explain each technique in more detail and illustrate how they overcome the verification obstacles.

3.5 Conclusion

In this chapter, first, the SCA-based verification of structurally simple multipliers was investigated. We provided a definition for structurally simple multipliers. Then, we chose three multipliers that match this definition and applied basic SCA-based verification to prove their correctness. The experimental results confirmed that SCA-based verification is fast and scalable when it comes to verifying these architectures.

Second, we studied the SCA-based verification of structurally complex multipliers. We presented a definition and explained two different ways that a structurally complex multiplier can be generated. Then, we chose three structurally complex multipliers and provided the experimental results for their verification. The explosion in the size of intermediate polynomials SP_i is a critical challenge that makes the verification of large structurally complex multipliers impossible.

Finally, we briefly introduced three techniques (i.e., local vanishing monomials removal, reverse engineering, and dynamic backward rewriting) to overcome the challenges of verifying structurally complex multipliers. In the next three chapters, we present each technique thoroughly and illustrate how they improve basic SCA-based verification.

Chapter 4
Local Vanishing Monomials Removal

Despite the success of SCA-based methods in the verification of structurally simple multipliers, verification of structurally complex multipliers is still a big challenge as an explosion happens in the number of monomials during backward rewriting. The dramatic increase in the number of monomials makes the calculations on intermediate polynomials very expensive and practically impossible in the case of large bug-free multipliers. A common understanding is that one of the main reasons for this explosion is the generation of redundant monomials known as vanishing monomials.

In this chapter, first, the problem of vanishing monomials is clarified using an example in the AIG representation. Then, we generalize the observations from the example and come up with the basic theory of vanishing monomials. Subsequently, the connection between vanishing monomials and the multiplier structure is investigated. Finally, we propose a technique to locally remove vanishing monomials and avoid the explosion during global backward rewriting. The proposed vanishing removal technique in this chapter has been published in [50, 53].

4.1 Introduction

SCA-based verification methods report very good results for structurally simple multipliers. However, they fail to prove the correctness of large structurally complex multipliers due to an explosion in the number of monomials during backward rewriting. One of the main reasons for this explosion is the generation of redundant monomials, known as vanishing monomials. These monomials are generated during the backward rewriting of structurally complex multipliers and reduced to zero after several steps of substitution. However, the huge number of vanishing monomials before cancellation causes a blow-up in computations.

A. Mahzoon et al., *Formal Verification of Structurally Complex Multipliers*,
https://doi.org/10.1007/978-3-031-24571-8_4

The concept of vanishing monomials was first introduced in [75]. The authors realized for the first time that SCA-based verification of integer multipliers containing parallel adders is unfeasible. They concluded that the main reason is vanishing monomials (monomials that always evaluate to zero), which appear in every algebraic model of these complex multipliers. They also proposed XOR rewriting scheme to eliminate vanishing monomials before they can cause a blow-up in computations.

Although the work of [75] is the first attempt to attack the vanishing monomials problem, it suffers from several shortcomings:

- It does not investigate the origin of vanishing monomials in structurally complex multipliers. Thus, it does not clarify why vanishing monomials appear and later get canceled out during backward rewriting.
- It does not show the relation between vanishing monomials and different multiplier architectures. Therefore, it remains unclear which stage of a multiplier and which architectures for that stage are responsible for the generation of vanishing monomials.
- The proposed approach works for some structurally complex multipliers. However, it fails for many others. Moreover, the verification run-times are long, and in practice, it cannot be applied to the multipliers bigger than 64×64.
- The approach is only applicable to the gate-level representation of a multiplier since the XOR gates should be explicitly available to perform XOR rewriting. Therefore, it cannot be used for the verification of AIG representations.

In this chapter, we overcome the shortcomings of [75] by (1) presenting the basic theory for the origin of vanishing monomials, (2) clarifying the relation between vanishing monomials and different multiplier structures, (3) proposing a fast technique to locally remove vanishing monomials, and (4) supporting multipliers in AIG representations since it gives us more flexibility in the detection of basic building blocks such as half-adders and full-adders.

4.2 Vanishing Monomials Example

As a circuit example, consider a 3×3 unsigned multiplier of type *simple partial product generator* $\circ$ *Wallace tree* $\circ$ *carry look-ahead adder* ($SP \circ WT \circ CL$). The AIG representation of the multiplier is shown in Fig. 4.1. Assume that half-adders and full-adders are identified before the backward rewriting process using reverse engineering techniques (more details in Chap. 5). This multiplier is structurally complex since the final stage is not only made of half-adders and full-adders. We use H_i and F_i to show the half-adder and full-adder blocks, respectively. The AIG nodes for H_4 (i.e., n_k, n_l, n_m, n_o), H_5 (i.e., n_x, n_y, n_z, n_t), and H_6 (i.e., n_p, n_u, n_q, n_r) are depicted in Fig. 4.1. For the rest of the half-adder and full-adder blocks, the internal nodes are not shown to keep the size of the example small and to avoid confusion. As can be seen in the figure, the inputs of the multiplier are $A = A_2A_1A_0$

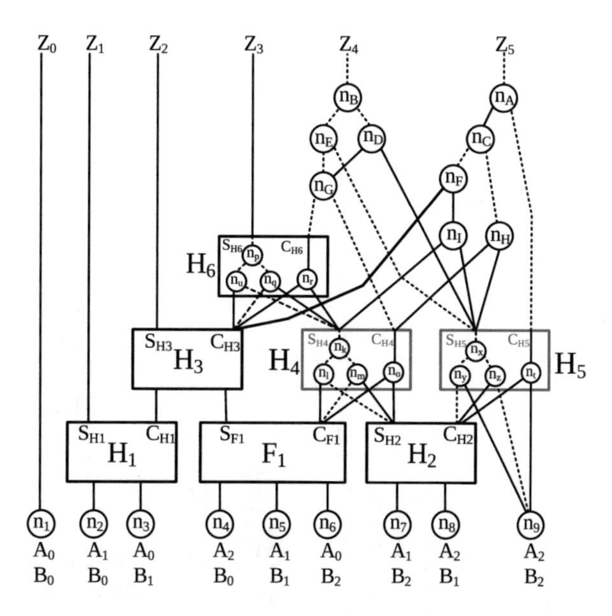

Fig. 4.1 AIG representation of a 3×3 structurally complex multiplier

and $B = B_2B_1B_0$ (to simplify the graph, we omit the input terminals but mark the successor nodes accordingly), while the output is $Z = Z_5Z_4Z_3Z_2Z_1Z_0$.

An excerpt of the substitution steps when performing backward rewriting for the 3×3 structurally complex multiplier is depicted in Fig. 4.2:

1. SP is the specification polynomial for the 3×3 multiplier at hand. Performing backward rewriting in the reverse topological order, i.e., substituting variables in SP with the node polynomials of Fig. 4.1, will finally result in the remainder zero since the considered AIG representation is correct.
2. In the first step of the backward rewriting, Z_5, which is one of the primary outputs of the circuit, is substituted with $1 - n_A$ (cf. NOT polynomial in Table 2.4). The result after the substitution is shown as the new polynomial SP_1. Since the coefficient of Z_5 is 32, we have to perform the multiplication $32(1 - n_A) = 32 - 32n_A$.
3. Then, n_A is substituted with $n_C - n_C C_{H_5}$ to obtain the new polynomial SP_2.
4. Subsequently, n_C is substituted with $1 - n_F - n_H + n_F n_H$ to obtain the new polynomial SP_3.
5. The next 13 steps of the backward rewriting are omitted.
6. Four intermediate backward rewriting steps, which are done by substituting n_o, S_{H_5}, n_z, and n_t, are presented, respectively. The intermediate result after the aforementioned substitutions is SP_{24}, cf. bold line.

As can be seen in Fig. 4.2, we have marked several monomials in red. These monomials are finally reduced to zero, i.e., after substituting S_{H_5}, C_{H_5}, n_x, n_y, n_z, and n_t, they are canceled out completely in SP_{24}. Hence, we call them vanishing monomials. Before explaining the origin and properties of vanishing monomials,

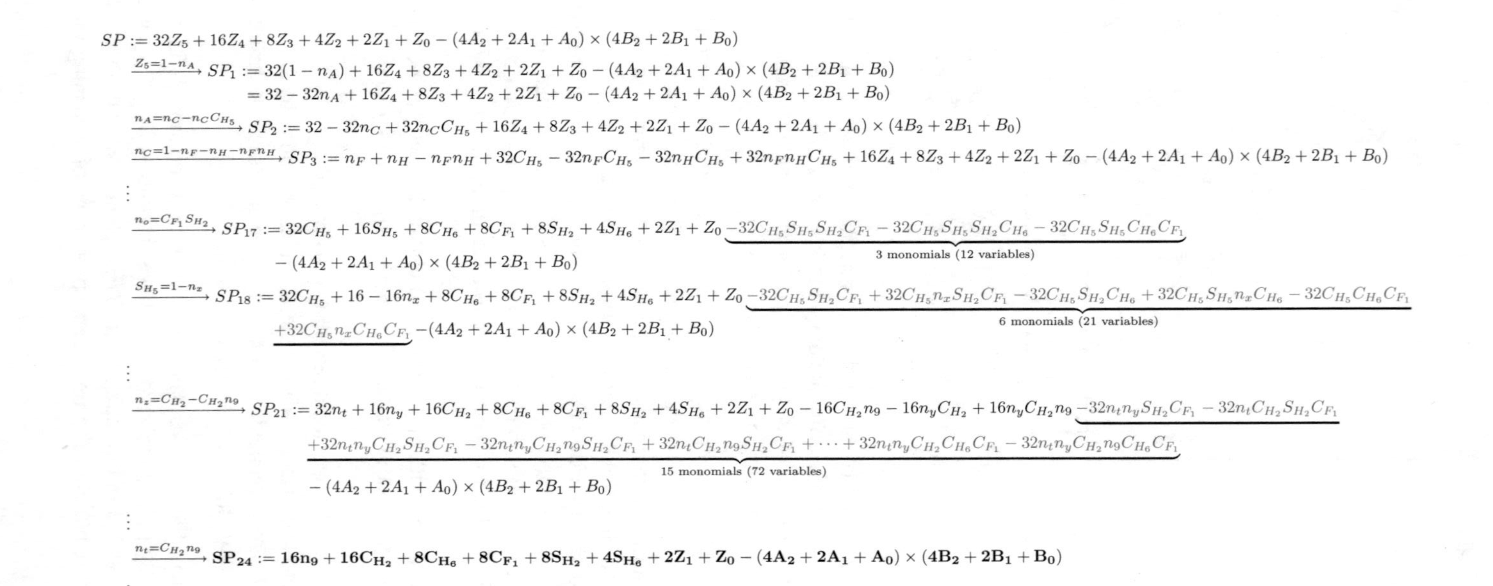

Fig. 4.2 Backward rewriting of a 3-bit $SP \circ WT \circ CL$ multiplier

we provide some numbers: there are 3 red monomials (12 variables) in SP_{17}: 6 red monomials (21 variables) in SP_{18}, and 15 red monomials (72 variables) in SP_{21}. These numbers show an explosion in the backward rewriting of the 3×3 structurally complex multiplier of Fig. 4.1. Please note that even more vanishing monomials appear in the complete backward rewriting steps. Now, two major questions arise:

1. Why are the red monomials finally reduced to zero in SP_{24}?
2. What is the origin of the red monomials?

For answering (1), just take a look at all 3 red monomials in SP_{17}. They all contain the product $C_{H_5}S_{H_5}$. In the next six substitution steps, this product is reduced to zero:

$$
\begin{aligned}
C_{H_5}S_{H_5} &= n_t(1-n_x) = n_t - n_t n_x \\
&= n_t - n_t(1 - n_y - n_z + n_y n_z) = n_t n_y + n_t n_z - n_t n_y n_z \\
&= C_{H_2}n_9(n_9 - C_{H_2}n_9) + C_{H_2}n_9(C_{H_2} - C_{H_2}n_9) \\
&\quad - C_{H_2}n_9(n_9 - C_{H_2}n_9)(C_{H_2} - C_{H_2}n_9) \\
&= \cancel{C_{H_2}n_9} - \cancel{C_{H_2}n_9} + \cancel{C_{H_2}n_9} - \cancel{C_{H_2}n_9} \\
&\quad - \cancel{C_{H_2}n_9} + \cancel{C_{H_2}n_9} + \cancel{C_{H_2}n_9} - \cancel{C_{H_2}n_9} = 0.
\end{aligned}
\tag{4.1}
$$

This is in line with the following observation: C_{H_5} and S_{H_5} are the outputs of H_5. As it is impossible to have both outputs of a half-adder "1" at the same time, the product $C_{H_5}S_{H_5}$ is always equal to zero. In summary, this is the reason why the red monomials finally vanish in SP_{24}.

Now, we give an answer to (2): as just discussed, all red monomials in SP_{14} contain the product $C_{H_5}S_{H_5}$. Traversing back all substitution steps (i.e., moving in the direction of the outputs on the AIG), this product originates from the product $n_C C_{H_5}$, formed via the substitution of $n_A = 1 - n_C C_{H_5}$ as can be seen in SP_1. Interpreting this observation on the AIG means that there are two paths[1] starting from the two half-adder outputs (here C_{H_5} and S_{H_5}) and these paths finally converge to a node (here n_A, node before output Z_5).

Overall, we conclude from this illustrating example that the origins of vanishing monomials are the AIG nodes where the half-adders' outputs converge, while the cancellation happens much later only after substituting the half-adders nodes' polynomials.

The next section provides the underlying theory of vanishing monomials. It also shows that vanishing monomials can be handled efficiently such that the size of intermediate polynomials SP_i does not grow dramatically during backward rewriting.

[1] Path 1: C_{H_5}, n_A; Path 2: $S_{H_5}, n_I, n_H, n_F, n_C, n_A$.

4.3 Basic Theory of Vanishing Monomials

We now generalize the observation from the illustrating example of the previous section. Therefore, we formulate the following theorem:

Theorem 4.1 *Assume that x and y are two AIG nodes representing the outputs of a half-adder. The product of x and y appears during the backward rewriting of a multiplier if at least one path from x and one path from y converge to an AIG node n_C, and the product of n_C inputs is not canceled out in calculations.*

Proof Figure 4.3a shows two paths starting from the half-adder outputs x and y and converging to the node n_C. The first path starting from x is a chain of AIG nodes $n_1, n_2, \ldots, n_i, n_C$. The second path starting from y consists of $n'_1, n'_2, \ldots, n'_j, n_C$. The edges connecting these nodes in the chains might be normal or complemented. Based on Table 2.4, we know that the polynomial of a 2-input AIG node contains the product of its inputs. Therefore, the polynomial of n_C can be written as

$$n_C = f(n_i, n'_j) + c\, n_i n'_j,$$

$$\begin{cases} n_C = n_i \wedge n'_j \rightarrow f(n_i, n'_j) = 0, \quad c = 1 \\ n_C = \neg n_i \wedge n'_j \rightarrow f(n_i, n'_j) = n'_j, \quad c = -1 \\ n_C = n_i \wedge \neg n'_j \rightarrow f(n_i, n'_j) = n_i, \quad c = -1 \\ n_C = \neg n_i \wedge \neg n'_j \rightarrow f(n_i, n'_j) = 1 - n_i - n'_j, \quad c = 1. \end{cases} \tag{4.2}$$

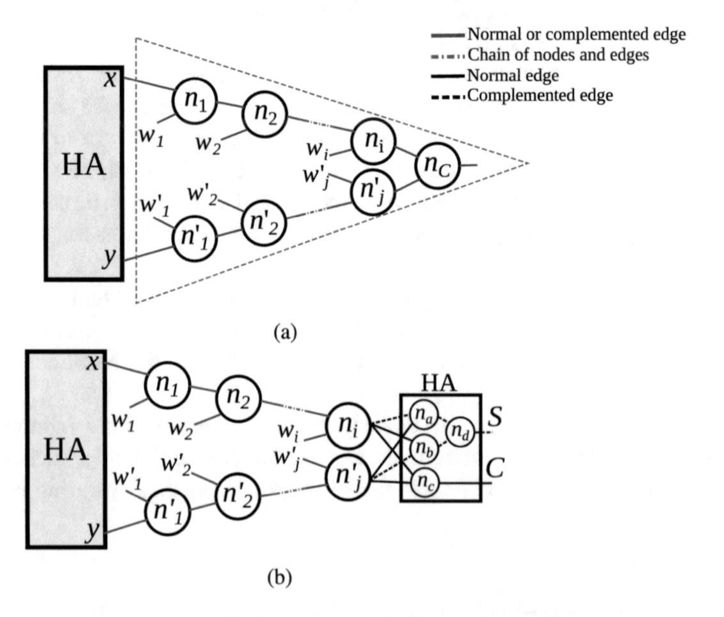

Fig. 4.3 Converging node cone. (**a**) General case. (**b**) Converging to a HA

The functions that describe n_i and n'_j based on the $x, y, w_1, w_2, \ldots, w_i, w'_1, w'_2, \ldots, w'_j$ are obtained after substituting the polynomials of the nodes located on the paths. After the substitutions, we get

$$\begin{aligned} n_i &= f'(w_1, w_2, \ldots, w_i, x), \\ n'_j &= f''(w'_1, w'_2, \ldots, w'_j, y). \end{aligned} \tag{4.3}$$

Based on Eqs. (4.2) and (4.3), we conclude

$$\begin{aligned} n_C &= f(n_i, n'_j) + cf'(w_1, w_2, \ldots, w_i, x) f''(w'_1, w'_2, \ldots, w'_j, y) \\ &= f(n_i, n'_j) + \underbrace{cxyT'_1 + cxyT'_2 + \cdots + cxyT'_r + cT_1 + cT_2 + \cdots + cT_s}_{\text{comes from the product} n_i n'_j}, \end{aligned} \tag{4.4}$$

where $cxyT'_h$ denotes the terms containing the product of x and y. Note that the xy product is generated as a result of multiplying two polynomials, one depending on x and the other one depending on y. After extracting the polynomial of n_C in Eq. (4.4), we now look at the backward rewriting process, i.e., we investigate the result of substituting n_C in the intermediate polynomial SP_i. Assume that the intermediate polynomial SP_i before substituting n_C with the node polynomial is as follows:

$$SP_i = n_C X'_1 + n_C X'_2 + \cdots + n_C X'_l + X_1 + X_2 + \cdots + X_q, \tag{4.5}$$

where $n_C X'_i$ denotes the terms containing n_C. Now, we distinguish between two cases that might happen after substituting n_C in Eq. (4.5):

1. *The product of n_C inputs, i.e., $n_i n'_j$, is canceled out early in calculations:* since the product $n_i n'_j$ is not contained in SP_i anymore, we can conclude that the product xy will not be generated in the next steps of backward rewriting. For example, assume that n_C is a part of another half-adder (see Fig. 4.3). Also assume that S and $2C$ are two terms in SP_i, while the rest of the terms are denoted by X_i. After substituting the polynomials of the half-adder nodes, the result is

$$\begin{aligned} SP_i &= S + 2C + X_i + \cdots + X_q, \\ SP_i \xrightarrow{S} SP_{i+1} &= 1 - n_d + 2C + X_i + \cdots + X_q, \\ SP_{i+1} \xrightarrow{C} SP_{i+2} &= 1 - n_d + 2n_c + X_i + \cdots + X_q, \end{aligned}$$

$$
\begin{aligned}
SP_{i+2} &\xrightarrow{n_d} SP_{i+3} = n_a + n_b - n_a n_b + 2n_c + X_i + \cdots + X_q, \\
SP_{i+3} &\xrightarrow{n_c} SP_{i+4} = n_a + n_b - n_a n_b + 2n_i n'_j + X_i + \cdots + X_q, \\
SP_{i+4} &\xrightarrow{n_a} SP_{i+5} = n'_j - n_i n'_j + n_b - n'_j n_b + n_i n'_j n_b + 2n_i n'_j \\
&\qquad + X_i + \cdots + X_q, \\
SP_{i+5} &\xrightarrow{n_b} SP_{i+6} = n'_j + \cancel{n_i n'_j} + n_i - \cancel{n_i n'_j} - \cancel{n_i n'_j} + \cancel{n_i n'_j} + \cancel{n_i n'_j} \\
&\qquad - \cancel{n_i n'_j} + X_i + \cdots + X_q \\
&\qquad = n_i + n'_j + X_i + \cdots + X_q.
\end{aligned}
\tag{4.6}
$$

as the product $n_i n'_j$ does not exist in SP_{i+5}, xy will not be generated later during backward rewriting.

2. *The product of n_C inputs, i.e., $n_i n'_j$, remains in calculations:* The product xy appears in the upcoming steps of backward rewriting as shown in Eq. (4.4).

□

Based on Theorem 4.1, we now make the following definitions, helping us with the early detection and cancellation of vanishing monomials.

Definition 4.1 Let n_C be an AIG node fulfilling Theorem 4.1, i.e., the outputs of at least one half-adder converge to n_C. Then, n_C is called a *converging node*.

Definition 4.2 Let n_C be a converging node. Then, the monomials containing the product of half-adder's outputs originating from n_C are *vanishing monomials*, as they are reduced to zero after substituting the half-adder nodes' polynomials.

For managing the size of intermediate polynomial SP_i during backward rewriting, it is essential to prevent the inclusion of vanishing monomials since for structurally complex multipliers an explosion occurs. Hence, the goal is to determine a vanishing-free polynomial representation for each converging node. In order to do this, we first look for the cones starting from half-adder outputs and ending in a converging node. Such a cone is called *Converging Node Cone* (CNC) in the rest of this book (see also the red area in Fig. 4.3a). Then, we locally extract the converging node polynomial based on the CNC inputs and remove all vanishing monomials. We use these polynomials during global backward rewriting. As a result, global backward rewriting becomes vanishing-free. Please note that local vanishing removal is independent of the circuit's function; thus, it is applicable to both correct and buggy multipliers.

Before explaining the algorithm to detect CNCs and remove vanishing monomials, we illustrate the relation between vanishing monomials and different multiplier architectures in the next section. In particular, we explain which multiplier stage is responsible for generating vanishing monomials. This helps us to narrow down the search space for finding CNCs and remove vanishing monomials efficiently.

4.4 Vanishing Monomials and Multiplier Architecture

Conducting several experiments on different multiplier architectures shows that vanishing monomials cause an explosion in the SCA-based verification of multipliers that use complex carry propagation hardware. This hardware is widely used in the third stage of multipliers (i.e., FSA) to reduce the propagation delay of trivial ripple carry adders [44]. As a result, at the cost of some growth in the area, the multiplier becomes faster. Carry look-ahead adder and parallel prefix adders (e.g., Kogge–Stone adder, Ladner–Fischer adder, and Han–Carlson adder) are among the architectures that use complex carry propagation hardware. As an illustrative example, we investigate the backward rewriting of a 4-bit carry look-ahead adder and show why many vanishing monomials are generated when it is used in the third stage of a multiplier.

Consider the Boolean formulation of a 4-bit carry look-ahead adder:

$$
\begin{aligned}
G_i &= x_i \wedge y_i, \\
P_i &= x_i \oplus y_i, \\
c_1 &= G_0 \vee (c_0 \wedge P_0), \\
c_2 &= G_1 \vee (G_0 \wedge P_1) \vee (c_0 \wedge P_0 \wedge P_1), \\
c_3 &= G_2 \vee (G_1 \wedge P_2) \vee (G_0 \wedge P_1 \wedge P_2) \vee (c_0 \wedge P_0 \wedge P_1 \wedge P_2), \\
c_4 &= G_3 \vee (G_2 \wedge P_3) \vee (G_1 \wedge P_2 \wedge P_3) \vee (G_0 \wedge P_1 \wedge P_2 \wedge P_3) \\
&\quad \vee (c_0 \wedge P_0 \wedge P_1 \wedge P_2 \wedge P_3),
\end{aligned}
\tag{4.7}
$$

where x_i and y_i are the ith bits of the first and second inputs, and c_i is the final carry. The AIG nodes dedicated to computing c_4 in the 4-bit carry look-ahead adder are shown in Fig. 4.4. If the Boolean formulation is transformed into the polynomial form, it consists of 31 monomials. However, 26 monomials contain the product of P_i and G_i. As P_i and G_i are the outputs of a half-adder, based on Eq. (4.1), their product equals zero (i.e., $P_iG_i = 0$). Therefore, all 26 monomials are reduced to zero after substituting the half-adder nodes' polynomials.

The generation of vanishing monomials during the substitution of node polynomials in the carry look-ahead adder can be justified by Theorem 4.1: P_i and G_i, which are outputs of a half-adder, converge to c_i, and the product of c_i inputs is not canceled out in the calculations. For example, in Fig. 4.4, the signal pairs (G_3, P_3), (G_2, P_2), and (G_1, P_1) converge to the node c_4. Thus, the product of P_i and G_i appears during backward rewriting.

The paths from half-adder outputs to converging nodes are usually long in multipliers using complex carry propagation hardware. Therefore, it takes several steps of substitution to reach the half-adder outputs during global backward rewriting. The generated vanishing monomials remain in the calculations in all these steps and cause an explosion in the number of monomials. As a result, finding CNCs

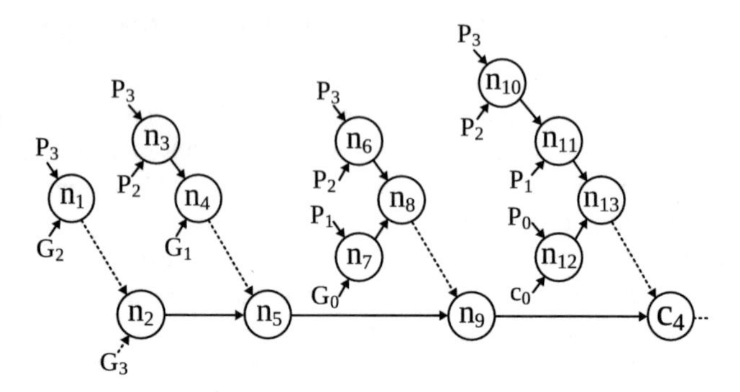

Fig. 4.4 Partial AIG representation of a 4-bit carry look-ahead adder

and obtaining vanishing-free polynomials for each converging node is essential to avoid explosion during global backward rewriting. However, detecting CNCs is impossible without identifying half-adders as all converging paths start from half-adder outputs. In Chap. 5, we explain how reverse engineering helps us to efficiently identify half-adders as well as other basic building blocks.

4.5 Removing Vanishing Monomials

After giving insight into the origins of vanishing monomials and their relation to multiplier architecture, we now propose a technique to remove them. First, we present an algorithm to detect CNCs in the AIG representation of a multiplier. Then, we show how the local removal of vanishing monomials in CNCs results in a vanishing-free global backward rewriting.

4.5.1 Converging Node Cone Detection

Algorithm 2 shows the proposed algorithm for detecting CNCs after reverse engineering. The algorithm receives the half-adders H and the extra nodes N as inputs and returns the set of CNCs as output. Extra nodes are nodes that are not part of any half-adder, full-adder, or compressors. As discussed in Sect. 4.4, the CNCs are the subsets of extra nodes. Therefore, we limit the search space for finding CNCs to these extra nodes. First, for each half-adder in H, all paths in N starting from the Sum and $Carry$ outputs are extracted (see Lines 3–4 in Algorithm 2). The end of a path is where the primary outputs or inputs of a basic building block (e.g., a half-adder or a full-adder) are reached. In fact, P_S and P_C contain all the possible node chains connecting the Sum and $Carry$ outputs of half-adders to the primary outputs or inputs of basic building blocks. Then, the paths in P_S and P_C

Algorithm 2 Detecting CNCs

Input: Set of half-adders H, Set of extra nodes N
Output: Set of converging node cones CN

1: $CN \leftarrow \emptyset$
2: **for each** $h \in H$ **do**
3: $P_S \leftarrow$ Find all paths starting from h_{Sum} in N
4: $P_C \leftarrow$ Find all paths starting from h_{Carry} in N
5: **for each** $p_S \in P_S$ **do**
6: **for each** $p_C \in P_C$ **do**
7: **if** $p_S \cap p_C \neq \emptyset$ **then**
8: g = First common member of p_S and p_C
9: $CNC \leftarrow [(p_S \cup p_C) - (p_S \cap p_C)] \cup \{g\}$
10: $CN \leftarrow CN \cup \{CNC\}$
11: $CN \leftarrow$ Merge all CNCs in CN with the same or included converging node
12: **return** CN

are checked to find out if there are paths which intersect (Lines 5–7). If that is the case, the first common member (i.e., g) is a converging node, as it is the first place where two paths from half-adder's outputs meet (Line 8). In order to determine the CNC for the corresponding converging node g, the union of two paths $(p_S \cup p_C)$ is subtracted by their intersection $(p_S \cap p_C)$ and g is added to the result to obtain all the nodes from half-adder's outputs to the converging node (Lines 9–10). This process is repeated for all half-adders to obtain the complete set of CNCs. Finally, all the cones with the same converging nodes (and thus the same outputs) and the cones whose converging nodes are included in other CNCs are merged, since there should be only one cone with a specific output signal (Line 11). In other words, cones C_1 and C_2 should be merged if (1) they have the same converging node or (2) converging node of C_1 (C_2), i.e., child cone, is a member of C_2 (C_1), i.e., parent cone. In the second case, the child cone is also returned as one of the outputs. In the local vanishing removal phase, the polynomials for the child cones are first calculated and their vanishing monomials are removed. These polynomials are later used in the calculation of the parent cone polynomial. Extracting the polynomials for child cones and removing their vanishing monomials help us to avoid the accumulation of vanishing monomials in parent cones during local backward rewriting.

Consider again the 3×3 structurally complex multiplier of Fig. 4.1. The two half-adders H_4 and H_5 are responsible for generating CNCs, as there are paths from the outputs of these half-adders converging to a node. Based on Algorithm 2, first the paths from the H_4 outputs are extracted: $p_1 = \{n_I, n_F, n_C, n_A\}$ and $p_2 = \{n_H, n_C, n_A\}$ are the paths starting from S_{H_4} and C_{H_4}, respectively. After calculating the intersection of these paths, we observe $p_1 \cap p_2 \neq \emptyset$. Thus, the first common member, i.e., n_C, is a converging node. Using the equation in Line 9 of Algorithm 2 results in the detection of the CNC $C_1 = \{n_C, n_H, n_F, n_I\}$. The members of a CNC are sorted based on the reverse topological order of the circuit. Hence, the first member of a CNC (n_C in C_1) is always the converging node. Additionally, $p_3 = \{n_r, n_G, n_E, n_B\}$ and $p_4 = \{n_G, n_E, n_B\}$ are two other paths starting from S_{H_4} and C_{H_4}. These paths converge to n_G; thus, after using Algorithm 2, we get $C_2 = \{n_G, n_r\}$. The complete list of detected CNCs after applying Algorithm 2 is $C_1 = \{n_C, n_H, n_F, n_I\}$, $C_2 = \{n_G, n_r\}$, $C_3 = \{n_A, n_C, n_F, n_I\}$, and $C_4 = \{n_A, n_C, n_H\}$, where C_1 and C_2 are related to H_4, and C_3 and C_4 are related to H_5. The cones C_3 and C_4 have the same converging node n_A; moreover, the converging node of C_1, i.e., n_C, is a member of C_3 and C_4. Consequently, we can merge these three cones to obtain $C = C_1 \cup C_3 \cup C_4 = \{n_A, n_C, n_H, n_F, n_I\}$. The cones C and C_2 as parent cones and the cone C_1 as a child cone are the final outputs of Algorithm 2.

The rest of the extra nodes which are not part of any basic building block or CNCs are grouped as *Fanout-Free Cones* (FFCs). In Fig. 4.1, $CF_1 = \{n_B, n_E, n_D\}$ is a fanout-free cone.

4.5.2 Local Removal of Vanishing Monomials

After detecting CNCs, the polynomial for each cone is extracted by a local backward rewriting. If a monomial that contains the product of half-adders outputs (i.e., vanishing monomial) appears during local backward rewriting, we remove the monomial instantly to avoid the generation of more vanishing monomials in the next steps. Finally, we have a set of polynomials that are completely vanishing-free. Note that if there are child cones in our CNCs list, their polynomials are first extracted. Then, they are used in the calculation of parent cone polynomials.

Consider $C = \{n_A, n_C, n_H, n_F, n_I\}$ which is a CNC of the multiplier in Fig. 4.1. It has a child node $C_1 = \{n_C, n_H, n_F, n_I\}$. The steps of local backward rewriting and the vanishing monomials removal for the child cone C_1 are as follows:

$$\begin{aligned} n_C &= 1 - n_F - n_H + n_F n_H \\ &= 1 - n_F - C_{H_4} S_{H_5} + n_F C_{H_4} S_{H_5} \\ &= 1 - n_I C_{H_3} - C_{H_4} S_{H_5} + n_I C_{H_3} C_{H_4} S_{H_5} \\ &= 1 - S_{H_4} S_{H_5} C_{H_3} - C_{H_4} S_{H_5} + \cancel{S_{H_4} C_{H_3} C_{H_4} S_{H_5}} \\ &= \boxed{1 - S_{H_4} S_{H_5} C_{H_3} - C_{H_4} S_{H_5}}. \end{aligned} \tag{4.8}$$

The child cone polynomial is used in extracting the polynomial for the parent cone C:

$$\begin{aligned} n_A &= n_C - n_C C_{H_5} \\ &= 1 - S_{H_4} S_{H_5} C_{H_3} - C_{H_4} S_{H_5} - C_{H_5} + \cancel{S_{H_4} S_{H_5} C_{H_3} C_{H_5}} + \cancel{C_{H_4} S_{H_5} C_{H_5}} \end{aligned}$$

$$= \boxed{1 - S_{H_4}S_{H_5}C_{H_3} - C_{H_4}S_{H_5} - C_{H_5}}. \quad (4.9)$$

The red monomials contain $S_{H_4}C_{H_4}$ or $S_{H_5}C_{H_5}$, which are the product of H_4 and H_5 outputs in Fig. 4.1, respectively. Therefore, they are canceled out immediately when they appear during local backward rewriting. Local removal of vanishing monomials in CNCs results in a vanishing-free global backward rewriting. Thus, we can avoid the explosion during the backward rewriting of structurally complex multipliers and overcome one of the SCA-based verification challenges.

4.6 Conclusion

In this chapter, first, an example of vanishing monomials generation and cancellation during the backward rewriting of a structurally complex multiplier was presented. This example clarifies how the generation of vanishing monomials can dramatically increase the size of intermediate polynomials. The observations are in line with the experimental results in Sect. 3.3 where an explosion occurred in the number of monomials during the verification of structurally complex multipliers.

Second, we came up with a theory for the origins of vanishing monomials. We proved that the origins of vanishing monomials are the AIG nodes where half-adders' outputs converge, while the cancellation happens much later after the substitution of half-adder nodes' polynomials. We also explained the relation between vanishing monomials and different multiplier architectures. We illustrated that vanishing monomials usually appear when a complex carry propagation hardware is used in the FSA (i.e., the third stage of a multiplier).

Finally, we proposed an algorithm to detect CNCs in the AIG representation of a multiplier. The local removal of vanishing monomials in CNCs results in a vanishing-free global backward rewriting. Thus, we can avoid the monomials explosion, caused by vanishing monomials, during the verification of structurally complex multipliers.

Chapter 5
Reverse Engineering

Integrating local vanishing monomials removal into the SCA-based verification is essential to avoid the explosion, caused by vanishing monomials, during the backward rewriting of structurally complex multipliers. However, it is important to identify basic building blocks, particularly half-adders, in advance in order to successfully detect CNCs and remove vanishing monomials. Moreover, the basic building blocks such as half-adders, full-adders, and compressors have a compact polynomial. The identification of these blocks and the utilization of their polynomials during backward rewriting speed up the whole verification process significantly.

In this chapter, first, the concept of atomic blocks is introduced. Then, the three main advantages of reverse engineering to identify atomic blocks before the SCA-based verification are discussed. Finally, we propose our reverse engineering technique to identify atomic blocks, including half-adders, full-adders, and (4:2) compressors, in the AIG representation of a multiplier. The proposed reverse engineering technique in this chapter has been published in [51, 53].

5.1 Introduction

The design hierarchy of a multiplier, including the boundaries of basic building blocks and their connections, is available at the register transfer level. However, this information is lost after synthesis. The gate-level netlist does not contain any design hierarchy, and the basic building blocks such as half-adders and full-adders are not explicitly visible. As a result, in the SCA-based verification of gate-level multipliers, the polynomials for each gate are captured and substituted in SP. The authors of [72, 94] first realized that the identification of half-adders and full-adders and using their polynomials during backward rewriting lead to a faster verification process. The reason is that half-adders and full-adders have very

A. Mahzoon et al., *Formal Verification of Structurally Complex Multipliers*,
https://doi.org/10.1007/978-3-031-24571-8_5

compact polynomials, which help to speed up the backward rewriting of multipliers. As a result, it became possible to verify a large structurally simple multiplier in just a few seconds. However, the works of [72, 94] suffer from some shortcomings:

- They only identify half-adders and full-adders and do not provide any technique to identify bigger building blocks such as compressors. Moreover, they do not clarify whether the identification of bigger building blocks can help speed up the verification process.
- They do not provide any vanishing monomials removal technique. Thus, they do not focus on an important advantage of identifying half-adders, i.e., detection of CNCs and removing vanishing monomials locally.

These shortcomings confine the application of the proposed methods to verifying structurally simple multipliers.

In this chapter, we overcome the shortcomings of [72, 94] by introducing the concept of atomic blocks and their importance in SCA-based verification. We clarify how the identification of bigger building blocks such as compressors in addition to half-adders and full-adders can help in the verification of some multiplier architectures. We also illustrate the role of atomic blocks in the detection of CNCs and successful vanishing monomials removal. As a result, for the first time, we extend the application of atomic blocks identification to the verification of structurally complex multipliers.

5.2 Atomic Blocks

In this section, we first define an atomic block. Then, we explain the structure and function of a (4:2) compressor that is a less-known atomic block compared to half-adders and full-adders.

Definition 5.1 An *atomic block* is a basic building block for a multiplier, which gets n one-bit binary inputs with the same bit positions,[1] and computes their sum as m one-bit binary outputs. The typical atomic blocks with 2, 3, and 5 inputs are half-adder, full-adder, and (4:2) compressor. The corresponding word-level relations are

$$\begin{aligned} HA(\text{in: } X, Y \quad \text{out: } C, S) \quad &\Rightarrow \quad 2C + S = X + Y, \\ FA(\text{in: } X, Y, Z \quad \text{out: } C, S) \quad &\Rightarrow \quad 2C + S = X + Y + Z, \\ CM(\text{in: } X, Y, Z, W, C_{in} \quad \text{out: } Co, C, S) \quad &\Rightarrow \quad 2Co + 2C + S \\ &= \quad X + Y + Z + W + C_{in}. \end{aligned} \tag{5.1}$$

[1] Assuming $A_{N-1}A_{N-2} \dots A_0$ and $B_{M-1}B_{M-2} \dots B_0$ as two binary numbers, A_i and B_i have the same bit positions.

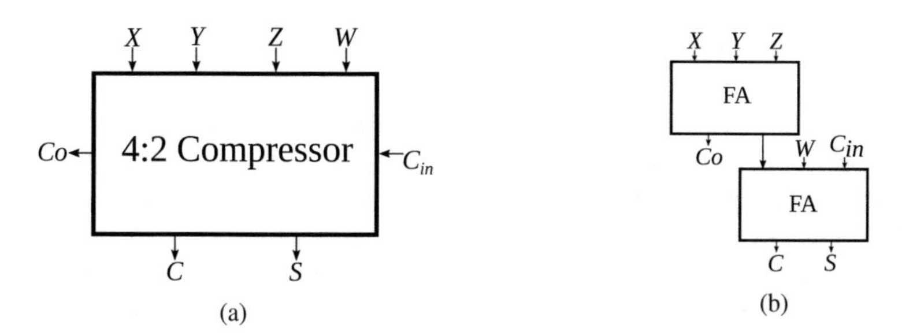

Fig. 5.1 (4:2) compressor. (**a**) Schematic. (**b**) Structure

Please note that this definition does not require a specific realization of an atomic block. In fact, only the respective mathematical relation is defined.

Compressors are used to reduce the delay in the PPA stage. Figure 5.1a shows the schematic of a (4:2) compressor. It has four inputs (X, Y, Z, W) along with an input carry (C_{in}) and two outputs (S, C) along with an output carry (Co). The output carry (Co) is usually connected to the input carry (C_{in}) of another compressor to create a compressor row. A (4:2) compressor can be constructed using two serially connected full-adders (see Fig. 5.1b). However, other optimized architectures have been proposed for a compressor to minimize the overall delay [44].

5.3 Advantages of Reverse Engineering in SCA

In this section, we explain the three main benefits of the atomic blocks identification in the SCA-based verification of multipliers.

5.3.1 Detecting Converging Node Cones

As discussed in Sect. 4.5, detecting CNCs and removing vanishing monomials are crucial to avoid explosion during backward rewriting. An important step before the CNC detection is the identification of all half-adders in the design, as a CNC always starts from half-adder outputs. It is also evident in Algorithm 2 (see Sect. 4.5) that the half-adders are the inputs of the algorithm.

The AIG representation of a half-adder is not unique. Using different atomic block libraries or applying optimization techniques results in half-adders with different node numbers. However, the function of a half-adder (i.e., the relation between outputs and inputs) never changes. Therefore, a reverse engineering technique to identify all atomic blocks (including half-adders) independent of their implementation is necessary to guarantee the detection of CNCs.

5.3.2 Limiting Search Space for Vanishing Removal

In Sect. 4.4, we explained that an explosion occurs during the backward rewriting of multipliers that use complex carry propagation hardware in their FSA. A part of the logic dedicated to this hardware always remains as the extra logic after reverse engineering. On the other hand, the second stage of a multiplier, which contains the largest number of AIG nodes, is fully made of atomic blocks. As a result, only a small part of the circuit that cannot be identified as atomic blocks, i.e., the extra logic, is responsible for generating vanishing monomials.

Figure 5.2a shows the ratio of atomic blocks logic to the entire logic in the different multiplier architectures after reverse engineering. Despite the fact that this ratio changes with respect to the design architecture, on average, atomic blocks constitute 70% of a multiplier. In addition, Fig. 5.2b depicts the atomic blocks ratio for the different PPA and FSA architectures. The results confirm that the PPA stage of many multipliers is completely made of atomic blocks. In contrast, the FSA stage of the multipliers using complex carry propagation hardware (e.g., BK, LF, and CL) is a mixture of atomic blocks and extra logic, and their ratio varies based on the architecture.

Overall, reverse engineering allows limiting the search space for finding converging gates to the extra logic in the FSA. This drastically reduces the search time in the local vanishing monomials removal phase.

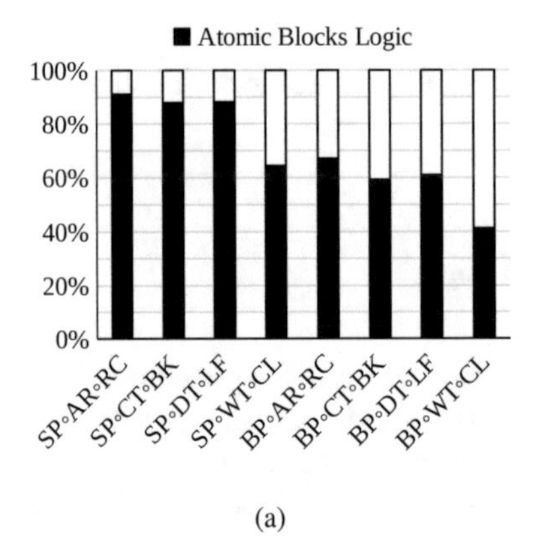

(a)

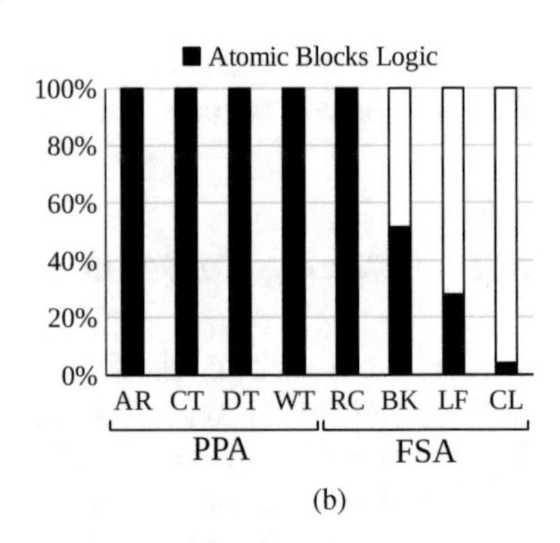

(b)

Fig. 5.2 Atomic blocks ratio in 64 × 64 multipliers. (**a**) Multiplier types. (**b**) Stage architectures. *SP* simple PPG, *BP* booth PPG, *AR* array, *CT* compressor tree, *WT* Wallace tree, *DT* Dadda tree, *RC* Ripple carry, *BK* Brent–Kung, *LF* Ladner–Fischer, *CL* Carry look ahead

5.3.3 Speeding up Global Backward Rewriting

As Eq. (5.1) indicates, there is always a compact algebraic relation between inputs and outputs of an atomic block, independent of their realization at the gate level. In general, this algebraic relation can be shown as

$$f(outputs) = g(inputs), \tag{5.2}$$

where $f(outputs)$ and $g(inputs)$ are functions based on the output and input signals, respectively. Therefore, if $f(outputs)$ appears in SP_i during backward rewriting, it can be substituted with $g(inputs)$ instantly. With respect to the fact that a large part of a design is constructed by atomic blocks (see Fig. 5.2a), detecting atomic blocks will speed up the global backward rewriting considerably.

As an example, assume C and S are the outputs, and X, Y, and Z are the inputs of a full-adder. If $2C + S$ appears in SP_i during backward rewriting, it can be substituted with $X + Y + Z$. As a result, we skip the substitution of the full-adder node polynomials and speed up the whole backward rewriting process.

Please note that if $f(outputs)$ does not explicitly appear, each output of an atomic block has to be substituted with its polynomial. The details of atomic blocks polynomial substitutions are explained in Chap. 6.

5.4 Proposed Reverse Engineering Technique

After clarifying the importance of atomic blocks in SCA-based verification, in this section, we propose our dedicated reverse engineering method to identify atomic blocks in multipliers.

At first, we collect the truth tables of atomic blocks in a library,[2] which has to be done only once. Then, we extract cuts in the AIG representation of a multiplier and check whether the output vector of each cut matches one of the output vectors in a truth table in our library. If we find a set of cuts with common inputs, whose output vectors match the truth table of an atomic block, we have identified the atomic block. Truth tables can be computed efficiently during cut enumeration for cuts with up to 16 inputs.[3] Moreover, they require an acceptable amount of memory. Thus, they are preferred to other symbolic representations such as BDDs [70, 90]. In the following, we explain the two steps of atomic blocks identification in detail.

[2] A truth table has several output vectors showing the value of the outputs for different input combinations, e.g., the truth table of a half-adder consists of two output vectors: one for *Sum* and another one for *Carry*.

[3] Cuts have a maximum of 5 inputs in our atomic block identification phase.

5.4.1 Atomic Blocks Library

First, we have to specify the mathematical functions of the atomic blocks and collect them in a library.

Assume that $f_i(x_1, x_2, \ldots, x_n)$ is the Boolean function for the ith output of an atomic block, and $x_1, x_2, \ldots, x_n$ are the atomic block inputs. The library for the atomic block should contain all the functions in the NPN class [40] of f_i, i.e., all the functions generated by swapping and complementing $x_1, x_2, \ldots, x_n$. Thus, the first step to create the library is the extraction of the NPN class for each atomic block output. Since the atomic block inputs are symmetric and have the same bit positions, swapping does not create new functions. As a result, the only transformation that leads to the generation of new functions for the NPN class is complementing. After extracting the NPN class, the truth table for each function in the class is stored in the library. By following this principle, the complete set of truth tables for half-adders and full-adders is obtained. We use the notation T_x to refer to the vector in column x of a truth table.

For example, the Boolean functions and the NPN classes for the two outputs of a half-adder are as follows:

$$Sum = X \oplus Y \xrightarrow{NPN\ class} X \oplus Y,\ \neg X \oplus Y,\ X \oplus \neg Y,\ \neg X \oplus \neg Y,$$

$$Carry = X \wedge Y \xrightarrow{NPN\ class} X \wedge Y,\ \neg X \wedge Y,\ X \wedge \neg Y,\ \neg X \wedge \neg Y. \tag{5.3}$$

If the first input of the half-adder is complemented (see Fig. 5.3a), the Boolean functions for the Sum and $Carry$ are $\neg X \oplus Y$ and $\neg X \wedge Y$, respectively. Therefore, the truth table contains two output vectors $T_{Sum} = 1001$ and $T_{Carry} = 0010$ based on Fig. 5.3b. On the other hand, if the both inputs are complemented (see Fig. 5.4a), the Boolean functions for the Sum and $Carry$ are $\neg X \oplus \neg Y$ and $\neg X \wedge \neg Y$, respectively. Thus, the output vectors are $T_{Sum} = 0110$ and $T_{Carry} = 0001$ based on Fig. 5.4b. Note that the complemented Boolean functions for each NPN class have to be considered as well since the outputs of an atomic block can be complemented. For example, it is possible that the $Carry$ output of a half-adder is complemented.

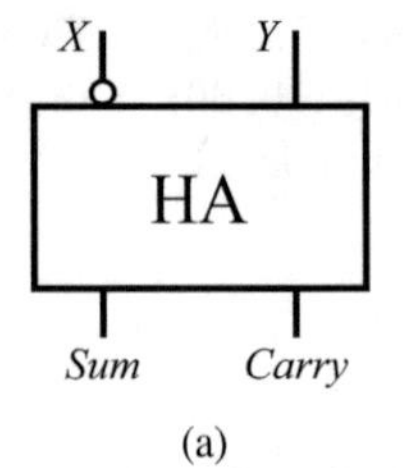

(a)

$T_{Sum} = 1001$ $T_{Carry} = 0010$

$\neg X$	Y	Sum	$Carry$
0	1	1	0
0	0	0	0
1	1	0	1
1	0	1	0

(b)

Fig. 5.3 Half-adder with one complemented input. (**a**) Structure. (**b**) Truth table

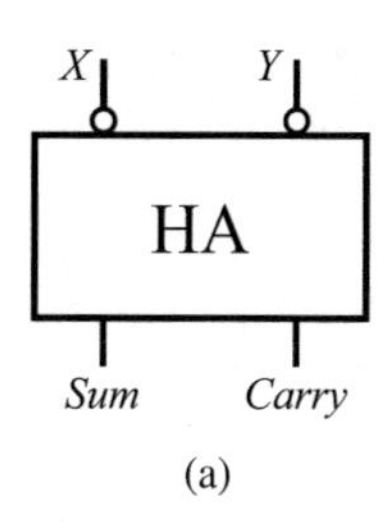

(a)

$T_{Sum} = 0110$ $T_{Carry} = 0001$

$\neg X$	$\neg Y$	*Sum*	*Carry*
0	0	0	0
0	1	1	0
1	0	1	0
1	1	0	1

(b)

Fig. 5.4 Half-adder with two complemented inputs. (**a**) Structure. (**b**) Truth table

The story for (4:2) compressor is different. The challenge originates from the fact that there are outputs with the same bit position. For example, this holds for the compressor of Fig. 5.1 with outputs S, C, and Co where C and Co have the same bit position (see CM word-level description in Eq. (5.1) where C and Co have the same coefficient). As a result, the value of these two outputs can be swapped for a certain input combination without changing the function of the compressor. This would lead to the generation of a large number of truth tables.

Table 5.1 shows the basic truth table (without complementing inputs) of a (4:2) compressor. Since C and Co have the same bit position, they can always be swapped. If the Boolean values of C and Co are not equal (red cells in Table 5.1), swapping them results in a completely new truth table. As in total there are 20 non-equal values of C and Co in the truth table of Table 5.1, $2^{20} = 1,048,576$ new truth tables are generated by swapping these values. To avoid dealing with millions of truth tables, we use the arbitrary values X_i in T_C, and its complement $\overline{X_i}$ in T_{Co}, where the ith value of T_C and T_{Co} is different. For example, in Table 5.1, T_C and T_{Co} can be encoded as follows:

$$T_C = 111X_3 1X_5X_6X_7 \ldots X_{24}X_{25}X_{26}0X_{28}000,$$
$$T_{Co} = 111\overline{X_3}1\overline{X_5X_6}X_7 \ldots \overline{X_{24}X_{25}X_{26}}0\overline{X_{28}}000. \quad (5.4)$$

The encoded values of T_C and T_{Co} in Eq. (5.4) cover all 2^{20} possible truth tables. In the next section, we explain how these encoded values are used in the identification of (4:2) compressors.

Finally, all the obtained truth tables of atomic blocks are stored in the *Atomic Blocks Library* (ABLib).

5.4.2 Atomic Blocks Identification

After creating ABLib, the next step is the identification of atomic blocks in the multiplier. Algorithm 3 presents the general algorithm for identifying atomic blocks with n inputs and m outputs using ABLib. The algorithm finds all cuts whose output

Table 5.1 (4:2) compressor truth table

X	Y	Z	W	C_{in}	S	C	C_O
1	1	1	1	1	1	1	1
1	1	1	1	0	0	1	1
1	1	1	0	1	0	1	1
1	1	1	0	0	1	1	0
1	1	0	1	1	0	1	1
1	1	0	1	0	1	1	0
1	1	0	0	1	1	1	0
1	1	0	0	0	0	1	0
1	0	1	1	1	0	1	1
1	0	1	1	0	1	1	0
1	0	1	0	1	1	1	0
1	0	1	0	0	0	1	0
1	0	0	1	1	1	1	0
1	0	0	1	0	0	1	0
1	0	0	0	1	0	1	0
1	0	0	0	0	1	0	0
0	1	1	1	1	0	1	1
0	1	1	1	0	1	1	0
0	1	1	0	1	1	1	0
0	1	1	0	0	0	1	0
0	1	0	1	1	1	1	0
0	1	0	1	0	0	1	0
0	1	0	0	1	0	1	0
0	1	0	0	0	1	0	0
0	0	1	1	1	1	1	0
0	0	1	1	0	0	1	0
0	0	1	0	1	0	1	0
0	0	1	0	0	1	0	0
0	0	0	1	1	0	1	0
0	0	0	1	0	1	0	0
0	0	0	0	1	1	0	0
0	0	0	0	0	0	0	0

vectors match one of the output vectors in a truth table in ABLib. If a set of cuts with common inputs has exactly the same output vectors as an atomic block truth table in ABLib, this set can be considered an atomic block. The inputs of the algorithm are the AIG G of a multiplier, the set of possible vectors for each output $ST_0, \ldots, ST_m$ from one concrete atomic block of ABLib, and the respective number of input bits n. The algorithm returns the list of found atomic blocks AB as output. First, all n-input cuts (cf. Definition 2.2) are computed on the AIG and stored in C (see Line 1). Then,

Algorithm 3 Atomic blocks identification

Input: Multiplier AIG G, Set of output vectors $ST_1, \ldots, ST_m$ from ABLib, Number of atomic block inputs n
Output: List of identified atomic blocks AB
1: $C \leftarrow$ FindCuts (G, n) ▷ Finding all n-input cuts
2: **for** $c_i \in C$ **do**
3: **for** $ST_j \in ST$ **do**
4: **if** TruthTable(c_i) $\in ST_j$ **then**
5: $PC_j = PC_j \cup c_i$
6: $SC \leftarrow$ Find the cuts with common inputs in $PC_0, PC_1, \ldots, PC_m$
7: $AB \leftarrow$ Merge the cuts with common inputs in SC
8: **return** AB

the output vectors of the cuts are checked to see whether there is a cut c_i whose output vector is the member of one of the output vector sets ST_j. If yes, i.e., the function of c_i is the same as the jth output of the atomic block, it is added to the list of possible candidates PC_j (Lines 2–5). Subsequently, the possible candidates are scanned to find the set of cuts with common inputs (Line 6). Finally, the cuts with common inputs are merged since we have found an atomic block (Line 7).

As an example, consider the 2×2 multiplier of Fig. 2.1b: $C_1 = \{n_5, n_6, n_8\}$ and $C_2 = \{n_7\}$ are among the extracted 2-input cuts. By computing the output vectors of these two cuts, it is realized that T_{C_1} and T_{C_2} are members of ST_S and ST_C that are the set of possible vectors for Sum and $Carry$ in ABLib, receptively. Moreover, C_1 and C_2 have the common inputs n_2 and n_3. Therefore, merging these two cuts results in identifying the atomic block $B = \{n_5, n_6, n_8, n_7\}$ that is a half-adder.

Please note that the output vectors C and Co of a (4:2) compressor (i.e., T_C and T_{Co}) have fixed values and encoded values. Thus, during the truth table matching (Lines 2–5 in Algorithm 3), we consider the encoded values as don't cares. Therefore, they could be either zero or one. Then, after finding the cuts with common inputs (Line 6), T_C and T_{Co} are checked to see whether they satisfy the conditions on the encoded values (see Eq. (5.4)), e.g., if X_i is encoded and its value is one in T_C, then X_i has to be zero in T_{Co}.

The run-time for computing cuts depends on the number of cut inputs, here n. In order to extract all atomic blocks efficiently, we first run Algorithm 3 for 2-input and 3-input cuts to detect all half-adders and full-adders. If the number of full-adders is less than 20% of the entire atomic blocks,[4] it can be concluded that the multiplier architecture has been implemented using larger atomic blocks, i.e., compressors. Hence, we run the algorithm for 5-input cuts to detect (4:2) compressors.

[4] This number has been justified by several experiments.

5.5 Conclusion

In this chapter, first, the definition of atomic blocks was presented. We introduced the three typical atomic blocks, i.e., half-adder, full-adder, and (4:2) compressor. We also gave details about the structure and function of a (4:2) compressor.

Second, the advantages of identifying atomic blocks before SCA-based verification were explained. Identification of half-adders that are the smallest atomic blocks is essential to detect CNCs and remove vanishing monomials. Moreover, it limits the search space for CNCs detection since a big part of the PPA and FSA stages is made of atomic blocks. In addition, using the atomic block polynomials speeds up global backward rewriting considerably.

Finally, our dedicated reverse engineering method to identify atomic blocks was proposed. It consists of two main steps: (1) collecting the truth tables of atomic blocks in a library, which has to be done only once, and (2) extracting cuts and checking whether their output vector matches one of the output vectors in a truth table in our library. We can identify all half-adders, full-adders, and (4:2) compressors in a multiplier using our reverse engineering technique.

Chapter 6
Dynamic Backward Rewriting

During global backward rewriting, there are usually several substitution candidates. Some of these substitutions do not increase the size of the intermediate polynomial considerably, while others lead to a blow-up in the number of monomials. It is particularly a big challenge for the verification of optimized and technology-mapped multipliers, where many atomic block boundaries have been removed. Thus, there are more substitution candidates at each step of backward rewriting. Finding the substitution candidates that keep the size of intermediate polynomial small at each step of backward rewriting is a crucial task to successfully verify optimized multipliers. It can be achieved by using a dynamic backward rewriting, which allows sorting substitution candidates during backward rewriting and restoring bad substitutions.

In this chapter, first, the challenges of backward rewriting for optimized multipliers are explained using an example. Then, our dynamic backward rewriting technique is proposed to overcome the challenges. This technique includes two main phases: sorting substitution candidates at each step of backward rewriting and restoring bad substitutions. The proposed dynamic backward rewriting technique in this chapter has been published in [54].

6.1 Introduction

The state-of-the-art SCA-based verification methods use a static backward rewriting to obtain the remainder. They set the order of substitutions before backward rewriting, based on, e.g., the distance of components (atomic blocks, cones, or gates) from inputs or outputs. The static substitution works successfully for structurally simple multipliers, since there is a limited number of choices for the polynomials substitutions at each step. Moreover, choosing different substitution orders does not change the size of intermediate polynomials significantly. However, the static

A. Mahzoon et al., *Formal Verification of Structurally Complex Multipliers*,
https://doi.org/10.1007/978-3-031-24571-8_6

substitution usually fails for structurally complex multipliers, particularly when the circuit is optimized. There are many possible substitution candidates at each step of backward rewriting. Some of these substitutions increase the size of intermediate polynomials drastically, while others do not change it significantly. Setting a static substitution order that keeps the size of intermediate polynomials small is an impossible task for dirty optimized multipliers.

While there have been some attempts to verify structurally complex multipliers with clean architectures (i.e., without optimization) [41, 75], the verification of optimized and technology-mapped multipliers is an unexplored area. Optimization destroys the clean boundaries between atomic blocks and multiplier stages. As a result, the substitution of atomic block polynomials is no more possible in many cases. The static backward rewriting, used in the SCA-based verification methods, fails for dirty optimized multipliers due to an explosion in the number of monomials. This creates major barriers for industrial use as most of the designs in the industry are area- and delay-optimized.

In this chapter, we come up with a new dynamic backward rewriting technique to overcome the challenges of verifying dirty multipliers after optimization. The proposed technique consists of two main steps:

- Sorting the substitution candidates in ascending order at each step of backward rewriting based on the number of their output occurrences in SP_i
- Restoring SP_i to its state before the substitution if a sharp increase happens in the number of monomials

The dynamic backward rewriting helps us to control the size of intermediate polynomials and avoid an explosion during the verification of optimized multipliers.

6.2 Multiplier Optimization

In this section, the effects of optimization on a multiplier architecture are explained. Then, we give insights into the obstacles of backward rewriting during the SCA-based verification of optimized multipliers.

6.2.1 Multiplier Structure After Optimization

In order to study the effects of optimization, we again consider the AIG of a 3×3 multiplier with $SP \circ AR \circ RC$ architecture before and after the optimization (see Fig. 3.7 in Sect. 3.3.1.2). We have used the proposed reverse engineering technique in Sect. 5.4 to detect the atomic blocks before and after optimization.

For the multiplier before the optimization (see Fig. 3.7a), the boundaries of atomic blocks and the connection between them are fully visible. However, after the optimization (see Fig. 3.7b), the boundaries of two full-adders F_1 and F_2 are destroyed. Thus, there are now several extra nodes in the second and third stages of the multiplier. These nodes were part of atomic blocks (i.e., full-adders) before optimization, but they cannot be identified as atomic blocks anymore.

6.2.2 *Backward Rewriting for Optimized Multipliers*

An important property of atomic blocks is the compactness of the polynomials describing their input/output relations. For example, the two most common atomic blocks, i.e., half-adder and full-adder, can be expressed using the following word-level relations:

$$\begin{aligned} HA(\text{in: } X, Y \quad \text{out: } C, S) \quad &\Rightarrow \quad 2C + S = X + Y, \\ FA(\text{in: } X, Y, Z \quad \text{out: } C, S) \quad &\Rightarrow \quad 2C + S = X + Y + Z. \end{aligned} \tag{6.1}$$

Let us have a look at backward rewriting when reaching atomic blocks and such word-level polynomials: Substituting an atomic block polynomial in SP_i only slightly changes the size of SP_i (i.e., the number of monomials). As a result, the size of SP_i remains almost identical.

In Fig. 3.7a, substituting the half-adder and full-adder polynomials increases the size of SP_i by zero or one, respectively. For example, substituting the F_3 polynomial $(2Out[5] + Out[4] = W_0 + W_1 + W_2)$ and the H_3 polynomial $(2W_0 + Out[3] = W_3 + W_4)$ in the first and second steps of backward rewriting results in[1]

$$\begin{aligned} &SP := 32Out[5] + 16Out[4] + 8Out[3] + \ldots \\ &SP \xrightarrow{F_3} SP_1 := 16W_2 + 16W_1 + 16W_0 + 8Out[3] + \ldots \\ &SP_1 \xrightarrow{H_3} SP_2 := 16W_2 + 16W_1 + 8W_3 + 8W_4 + \ldots \end{aligned} \tag{6.2}$$

On the other hand, substituting later $n_0, n_1, \ldots, n_8$ (AND gates) polynomials results in the cancelation of two terms per AND gate and thus reduces the size of SP_i by two per AND gate.

If we now look at the overall backward rewriting algorithm (and not only at a single substitution step), all existing SCA-verifiers use a static pre-determined substitution order derived from the topological sorting of the AIG nodes/circuit elements. This worked out fine so far, as sharp increases in the size of SP_i are not occurring if most of the time atomic blocks are handled as a whole—either by substituting word-level polynomials (if possible) or at least by using a substitution order that substitutes polynomials for the single outputs of atomic blocks strictly

[1] The case that the left-hand sides of word-level relations for atomic blocks do not occur in the needed form in the SP_i is discussed later in Sect. 6.3.

one after the other. If we lose the block boundaries in optimized multipliers, the exploitation of this knowledge is not possible anymore. Thus, we observe much larger polynomials and in the worst case a blow-up.

In Fig. 3.7b, after substituting the F_3 and H_3 polynomials, there are several degrees of freedom for the subsequent substitutions respecting the reverse topological order of the $n_0, n_1, \ldots, n_{24}$ nodes and H_1, H_2 components. A straightforward topological order results in a sudden increase in the SP_i size, and even an explosion in larger multipliers. For example, using a topological order SO_1 during the backward rewriting of an optimized 8×8 array multiplier results in an intermediate polynomial with 106, 938 monomials. However, by using a different order SO_2, we can limit the maximum size of SP_i to 203.

6.3 Proposed Dynamic Backward Rewriting Technique

In this section, we first define some useful terms. Then, we propose our algorithm for dynamic backward rewriting, which allows verification of optimized multipliers. The algorithm consists of two important phases: sorting substitution candidates and restoring bad substitutions.

6.3.1 Definitions

Before introducing the dynamic backward rewriting, we first make several definitions:

Definition 6.1 Atomic blocks, CNCs, and FFCs are called *Components*. A component has one output if it is a CNC or an FFC, and may have several outputs if it is an atomic block. A multiplier consists of several components.

Definition 6.2 A component has a polynomial describing the (word-level) relation of output(s) and inputs. In components with one output (i.e., CNCs and FFCs), we have

$$Out = F(IN_1, IN_2, \ldots, IN_n), \tag{6.3}$$

where Out is the sole output of the component, and $F(IN_1, IN_2, \ldots, IN_n)$ is a polynomial based on the component inputs. In components with more than one output (i.e., atomic blocks), we have

$$\begin{aligned}
Out_1 &= F_1(IN_1, IN_2, \ldots, IN_n), \\
Out_2 &= F_2(IN_1, IN_2, \ldots, IN_n), \\
&\ldots
\end{aligned}$$

$$Out_m = F_m(IN_1, IN_2, \ldots, IN_n), \tag{6.4}$$

where each output can be described as a polynomial based on the primary inputs. There is also a more compact relation between outputs and inputs in these components:

$$G(Out_1, Out_2, \ldots, Out_m) = F(IN_1, IN_2, \ldots, IN_n), \tag{6.5}$$

where G and F are polynomials based on the outputs and inputs, respectively.

For example, in a full-adder with the inputs X, Y, and Z, and the outputs C and S, we have

$$C = XY + XZ + YZ - 2XYZ, \tag{6.6}$$

$$S = X + Y + Z - 2XY - 2XZ - 2YZ + 4XYZ, \tag{6.7}$$

$$2C + S = X + Y + Z, \tag{6.8}$$

where Eqs. (6.6) and (6.7) show the polynomials for carry and sum outputs, respectively. However, Eq. (6.8) indicates the compact relation between outputs and inputs.

Definition 6.3 Substituting a component polynomial in the intermediate specification polynomial (i.e., SP_i) means finding and replacing all the occurrences of component output(s) with the corresponding polynomial(s) in SP_i.

To ensure correct substitution, we need two rules:

- In components with more than one output (i.e., atomic blocks), which have a compact word-level relation between inputs and outputs (see Eq. (6.5))), we first search for $G(Out_1, Out_2, \ldots, Out_m)$ in SP_i. Then, we have to distinguish between two cases: If we have found $G(Out_1, Out_2, \ldots, Out_m)$, we directly substitute it with $F(IN_1, IN_2, \ldots, IN_n)$; otherwise, we substitute each output with the corresponding polynomials. Finding the exact G polynomial is sometimes not possible, particularly in optimized multipliers. Therefore, following this rule guarantees the correct substitution.
- Assume that C is a component with output W, and $C_1, C_2, \ldots, C_n$ are components having W as one of their inputs (see Fig. 6.1). The C polynomial has to be substituted only after substituting the $C_1, C_2, \ldots, C_n$ polynomials. Following this rule guarantees that a component polynomial needs to be substituted only once during the backward rewriting as for example the signal W never appears again in the SP_i after substituting the C polynomial. A component that satisfies this rule and is ready for substitution is called a *Candidate*.

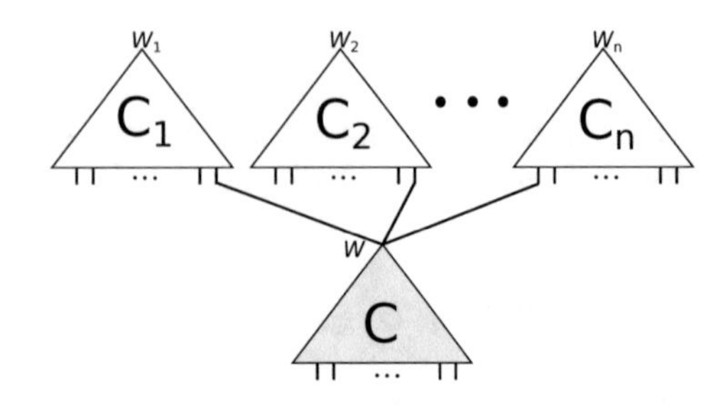

Fig. 6.1 Substitution candidates

Algorithm 4 Dynamic backward rewriting

Input: Specification Poly SP, Set of Components B, Set of Component Polys P
Output: Remainder r

1: $SP_i \leftarrow SP$
2: **while** B is not empty **do**
3: $\quad CB \leftarrow$ Find eligible candidates in B for substitution
4: $\quad$ **for each** $b \in CB$ **do**
5: $\quad\quad O[b] \leftarrow$count_occurrence($b$, SP_i)
6: $\quad sortedCB \leftarrow$ Sort CB in ascending order based on O
7: $\quad SP_{old} \leftarrow SP_i$
8: $\quad threshold \leftarrow 0.1$; $j \leftarrow 0$; $no\_candidate \leftarrow TRUE$
9: $\quad$ **while** no_candidate is TRUE **do**
10: $\quad\quad SP_i \leftarrow$ Substitute $(P_{sortedCB[j]}, SP_i)$
11: $\quad\quad$ **if** $(size(SP_i) - size(SP_{old}))/size(SP_{old}) < threshold$ **then**
12: $\quad\quad\quad$ Remove $sortedCB[j]$ from B
13: $\quad\quad\quad no\_candidate \leftarrow FALSE$
14: $\quad\quad$ **else**
15: $\quad\quad\quad SP_i \leftarrow SP_{old}$; $j \leftarrow j + 1$
16: $\quad\quad\quad$ **if** $j \geq size(SortedCB)$ **then**
17: $\quad\quad\quad\quad j \leftarrow 0$; $threshold \leftarrow threshold \times 2$
18: $r \leftarrow SP_i$
19: **return** r

6.3.2 Algorithm

There are several substitution candidates at each step of backward rewriting. As we discussed in Sect. 6.2.2, choosing a substitution order that keeps the size of SP_i as small as possible and avoids explosion in optimized multipliers is of utmost importance. Therefore, we propose dynamic backward rewriting.

Algorithm 4 shows our proposed dynamic backward rewriting. The algorithm receives the set of components and the specification polynomial as inputs and performs all possible backward rewriting steps. First, the substitution candidates are found (see Line 3 in Algorithm 4). Then, the number of times which the output of each candidate occurs in SP_i is counted (Lines 4–5). We sort the candidates in ascending order based on the number of their output occurrences in SP_i (Line 6), and basically, we try to substitute them in SP_i according to that order. The following example motivates this choice:

Consider the polynomial $P = a + 4abc - 2ad - 2adc$. Substituting a variable occurring k times in P by a polynomial containing h monomials increases the number of monomials in the result by $k \cdot (h - 1)$ in the worst case. For example, substituting $a = x + y + z + xz$ in P results in a new polynomial with 16 monomials. For this reason, we refrain from substituting variables with large numbers of occurrences in P first. Instead, we start with variables having smaller numbers of occurrences and hope that cancelation of monomials improves the situation when we finally arrive at the substitution of later variables. Assume in the example that we substitute $b = xy$ first, then $c = xz$ and then $d = xyz$, leading to $P = a$. After substituting $a = x + y + z + xz$, we arrive at the final result of the series of substitutions without exceeding a number of four monomials in between.

In our experiments, we observed that the order mentioned above successfully keeps the sizes of intermediate polynomials small in many cases. Nevertheless, this method is only heuristic and may fail. Therefore, before each substitution, we make a copy of SP_i (Line 7) to which we may backtrack if the size of the polynomial grows too much. Altogether, this leads to a *dynamic* substitution order based on the growth behavior we actually observe.

As an example, consider the polynomial $P = abx + aby - 2abxy + ab + a$. Substituting the 4 occurrences of b by $m + n - mn$ leads to a polynomial with 13 monomials. After substituting $a = xy$ in the resulting polynomial, we arrive at 4 monomials. However, if we substitute a (with one occurrence more than b) first, we obtain 2 monomials after the first substitution and again 4 monomials after substituting b. Of course, we prefer the second order which we obtain if we discard the substitution of b due to the size of the intermediate result.

In Algorithm 4, this idea is implemented as follows: After making a copy of SP_i (Line 7), the polynomial for the first component in the sorted candidate list is substituted in SP_i (Line 10). We check the increase in the size of the intermediate polynomial after the substitution. To avoid a sharp increase in the number of monomials, we set the increase threshold to 10% (Line 8). If the increase is less than the threshold, we recognize it as a successful substitution, i.e.,

$$\frac{size(SP_i) - size(SP_{old})}{size(SP_{old})} < threshold. \tag{6.9}$$

Thus, the candidate is removed from the list of components (Lines 12–13), and the process is repeated by finding a new set of candidates (Line 3). On the other hand, if the increase is more than the threshold, then SP_i is restored to its state before substitution, and the process continues by substituting the next candidate in the sorted list (Line 15). If there is no substitution of any candidate that satisfies the threshold limit, the value of the threshold is multiplied by two, and the process is repeated from the first candidate in the sorted list (Lines 16–17).

Please note that the depth of backtracking is one in Algorithm 4. It is possible to obtain smaller intermediate polynomials if we increase the depth, i.e., if the depth is d, SP_i can be restored up to d steps. However, backtracking more than one step is very costly in terms of run-time and memory usage, since we have to always keep the copies of all intermediate polynomials from step i to $i + d$, i.e., $SP_i, SP_{i+1}, \ldots, SP_{i+d}$. These intermediate polynomials are usually big, particularly for large optimized multipliers; thus, making the copies and storing them consume considerable time and memory.

6.4 Conclusion

In this chapter, first, the effects of optimization on the multipliers structure and the process of global backward rewriting were explained. In general, optimization removes the boundaries of several atomic blocks. As a result, the static backward rewriting does not work anymore since there are many substitution candidates, and most of them increase the size of SP_i dramatically. We need to choose a candidate at each step of backward rewriting in such a way that the size of SP_i remains small throughout the whole verification process.

Second, we proposed our dynamic backward rewriting technique to overcome the challenges. Our technique consists of two main phases that take place at each step of backward rewriting. In the first phase, the substitution candidates are sorted ascendingly based on the number of their output occurrences in SP_i. Then, in the second phase, it is checked whether a sharp increase happens in the number of monomials after the substitution. If it is the case, SP_i is restored to its state before the substitution, and we continue with the next candidate.

In the next chapter, we show that our proposed dynamic backward rewriting works for various optimized multipliers. However, if the optimization is very harsh, i.e., it removes the boundaries of most atomic blocks, the proposed method might fail. Increasing the depth of backtracking while finding a way to keep the run-time and memory usage small could be a possible solution for the verification of highly optimized multipliers.

Chapter 7
SCA-Based Verifier RevSCA-2.0

SCA-based verification reports very good results when it comes to proving the correctness of structurally simple multipliers. However, it fails to verify structurally complex multipliers as an explosion happens in the number of monomials during backward rewriting. The basic SCA-based verification can be improved by the integration of three techniques (i.e., reverse engineering, local vanishing monomials removal, and dynamic backward rewriting), helping us to attack the hard problem of verifying structurally complex multipliers.

In this chapter, we propose our new SCA-based verifier RevSCA-2.0. RevSCA-2.0 uses SCA as its core verification engine, but it takes advantage of the three introduced techniques in this book to improve and extend SCA for the verification of structurally complex multipliers. We first present the top-level overview of our verification approach and explain how the three techniques are integrated into SCA-based verification. Then, the details of the RevSCA-2.0 implementation are given. Subsequently, we introduce our multiplier generator GenMul, which helps us with the generation of large multiplier benchmarks. Finally, we report the verification results for a wide variety of structurally complex multipliers, including clean and dirty optimized architectures. Experimental results confirm the efficiency of our approach in the verification of various benchmarks, including industrial multipliers. Parts of this chapter have been published in [52, 53].

7.1 Introduction

Over the last few years, several SCA-based verification methods have been proposed to verify integer multipliers. These methods use the basic SCA-based verification introduced in Sect. 2.5.3 as their core engine, but they also try to improve the run-time, memory usage, or applicability of the SCA-based method by integrating some novel techniques. Despite the progress in the formal verification of structurally

A. Mahzoon et al., *Formal Verification of Structurally Complex Multipliers*,
https://doi.org/10.1007/978-3-031-24571-8_7

simple multipliers, the research works dedicated to the SCA-based verification of structurally complex multipliers are very few. The authors of [75] proposed an XOR rewriting technique to remove vanishing monomials before global backward rewriting. The method only works for some structurally complex multipliers, and it fails for many others. The proposed method in [41] combines SAT and SCA to verify structurally complex multipliers. The approach reports good results for a wide variety of multipliers with clean architectures. However, it is not robust against optimization, and it fails for almost all dirty architectures after optimization or technology mapping.

In this chapter, we propose our SCA-based verifier RevSCA-2.0 that integrates three introduced techniques in the previous chapters into SCA-based verification. The techniques are:

- Reverse engineering: Identification of atomic blocks, including half-adders, full-adders, and (4:2) compressors, is essential to detect CNCs and remove vanishing monomials. It also speeds up global backward rewriting.
- Local vanishing monomials removal: Early detection and removal of vanishing monomials are crucial to avoid an explosion in the number of monomials during the verification of structurally complex multipliers.
- Dynamic backward rewriting: Finding the substitution candidates that do not increase the size of intermediate polynomials dramatically at each step of backward rewriting and restoring bad substitutions are necessary for the verification of structurally complex multipliers, particularly dirty optimized architectures.

RevSCA-2.0 can be used to verify structurally complex multipliers, including both clean and dirty optimized architectures. In order to show the efficiency of RevSCA-2.0 in the verification of various multipliers, we need a big set of benchmarks containing a wide variety of architectures with different sizes. However, the available benchmarks are usually small, i.e., their input sizes are limited to 64×64. In order to overcome this obstacle, we propose our multiplier generator GenMul. GenMul generates a large variety of structurally complex multipliers with arbitrary input sizes; thus, we can challenge RevSCA-2.0 with very large benchmarks.

The experimental results in this chapter confirm that RevSCA-2.0 verifies a large set of structurally complex multipliers with up to several million gates, while the other state-of-the-art SCA-based verification methods can only support a limited number of benchmarks.

7.2 Top-Level Overview

In this section, we present the top-level overview of our SCA-based verifier RevSCA-2.0. We explain how we can integrate the three proposed techniques in the previous chapters into SCA-based verification in order to successfully prove the correctness of structurally complex multipliers.

Algorithm 5 REVSCA-2.0

Input: Multiplier AIG G
Output: TRUE if the circuit is correct, and FALSE otherwise
1: $SP \leftarrow$ CreateSP(G)
2: AB, $N \leftarrow$ ReverseEngineering(G) ▷ AB is the set of atomic blocks, N is the set of extra nodes
3: $CN \leftarrow$ FindCNCs(N, filter_HAs(AB)) ▷ CN is the set of CNCs
4: $CF \leftarrow$ FindFFCs(N, CN) ▷ CF is the set of FFCs
5: $C \leftarrow CN \cup CF$
6: $F \leftarrow$ ExtractVanishingFreePolys(C) ▷ F is the set of cone polynomials
7: $r \leftarrow$ DynamicBackwardRewriting(SP, F, AB) ▷ r is the remainder
8: **if** $r == 0$ **then**
9: **return** *TRUE*
10: **else**
11: **return** *FALSE*

In our proposed method, first, SP is generated. For an $N \times N$ unsigned multiplier, we always have

$$SP = \sum_{i=0}^{2N-1} 2^i Z_i - \left(\sum_{i=0}^{N-1} 2^i A_i\right) \times \left(\sum_{i=0}^{N-1} 2^i B_i\right). \tag{7.1}$$

On the other hand, for an $N \times N$ signed multiplier, we have

$$\begin{aligned} SP = {} & -2^{2N-1} Z_{2N-1} + \sum_{i=0}^{2N-2} 2^i Z_i \\ & - \left(-2^{N-1} A_{N-1} + \sum_{i=0}^{N-2} 2^i A_i\right) \times \left(-2^{N-1} B_{N-1} + \sum_{i=0}^{N-2} 2^i B_i\right), \end{aligned} \tag{7.2}$$

where $A_{N-1}A_{N-2}\ldots A_0$ and $B_{N-1}B_{N-2}\ldots B_0$ are the inputs and $Z_{2N-1}Z_{2N-2}\ldots Z_0$ is the output.

Then, all atomic blocks, including half-adder, full-adders, and (4:2) compressors, are identified using the proposed reverse engineering algorithm (see Algorithm 3 in Sect. 5.4.2). Subsequently, all CNCs starting from the half-adder outputs are detected using Algorithm 2 in Sect. 4.5, and the polynomial for each CNC is extracted by the substitution of the node polynomials in the cone. The CNC polynomial determines the output of the cone (i.e., output of the converging node) based on its inputs. We know that a vanishing monomial contains the product of half-adder outputs, and these outputs are the inputs of CNCs. Therefore, the vanishing monomials appear in the extracted CNC polynomials. Local removal of vanishing monomials from these polynomials leads to a set of vanishing-free polynomials. Finally, the proposed dynamic backward rewriting technique in Algorithm 4 (see Sect. 6.3) is performed to obtain the remainder.

Algorithm 5 shows the pseudocode of REVSCA-2.0. In the first step, SP is created based on the input and output bit-width of the multiplier (Line 1). Then, atomic blocks are identified using the proposed reverse engineering technique (Line 2). The CNCs are extracted based on the identified HAs (filter_HAs(AB)) and the set of extra nodes from the reverse engineering phase (Line 3). The rest of the nodes that are not part of any atomic blocks or CNCs are grouped based on the fanout-free regions to create the set of FFCs. In the next step, the polynomial for each cone is extracted by substitution, and vanishing monomials are locally removed (Line 6). Finally, dynamic backward rewriting is performed to obtain the remainder (Line 7). If the resulting remainder equals zero, the circuit is correct; otherwise, it is buggy (Lines 8–11).

In the next section, we explain the details of the REVSCA-2.0 implementation, including data structures and used libraries for reverse engineering.

7.3 Implementation

In this section, first, the polynomial data structure used in the implementation of our SCA-based verifier REVSCA-2.0 is explained. Then, we give insight into reverse engineering and the used libraries.

7.3.1 Polynomial Data Structures

A big part of computations during the SCA-based verification of integer multipliers is related to polynomials. Two of the most important computations are as follows:

- Substitution of a variable v with a polynomial f in P. In this case, we need to first search for all occurrences of v in P; then, we remove v and multiply out the monomials containing v by f to get the new polynomial. For example, the substitution of c with the polynomial $f = a+b$ in $P = abc+2cd-3ab$ results in

$$P_1 = ab(a+b) + 2(a+b)d - 3ab = ab + ab + 2ad + 2bd - 3ab. \quad (7.3)$$

- Searching for all terms with the same monomials after substitution and adding up their coefficients. Since simplifications and cancelations usually happen during backward rewriting and after each substitution, looking for terms with the same monomials and adding up their coefficient are crucial. For instance, the polynomial P after the substitution has three terms with the same monomial, i.e., ab. Thus, their coefficients have to be added up to obtain the simplified polynomial:

$$P_1 = ab + ab + 2ad + 2bd - 3ab = -ab + 2ad + 2bd. \quad (7.4)$$

Table 7.1 Hash table for the polynomial $P = abc + 2cd - 3ab$

Keys	Buckets
{1,2,3}	1
{3,4}	2
{1,2}	-3

In order to efficiently store polynomials and perform computations on them, we need a suitable data structure. The data structure should allow us to carry out both aforementioned computations, i.e., substitution and simplification, as fast as possible at each step of backward rewriting.

We propose a polynomial data structure based on the *hash table*. Our data structure has the following properties:

- The keys in the hash table are monomials represented as a set of positive integer values. Each value in the set is corresponded to a variable in the monomial. Since a monomial never has duplicate variables, the members of the set are unique.
- The buckets in the hash table are coefficients represented with the integer values.
- The hash table maps each key (i.e., monomial) to a bucket (i.e., coefficient); therefore, it represents a polynomial.

Table 7.1 shows the hash table for the polynomial $P = abc + 2cd - 3ab$. The variables a, b, c, and d are represented with the positive integer numbers 1, 2, 3, and 4, respectively. Each monomial is now a key in the hash table, and it is mapped to a coefficient.

The hash table representation of a polynomial brings us a big advantage during backward rewriting: The keys in a hash table are unique, i.e., after substitution, simplification is carried out automatically without requiring extra computations. We only need to add up the coefficients of generated monomials after substitution to the buckets with respect to the corresponding keys. We add the key to the hash table if it already does not exist. For example, the polynomial P_1 in Eq. (7.4) has to be simplified after the substitution. There is already a key for ab monomial in the hash table; thus, we only need to add the coefficients for newly generated ab monomials to the bucket, i.e., $-3 + 1 + 1 = -1$. On the other hand, there are no keys for the monomials ad and bd; therefore, we need to create the keys. Please note that the old keys that do not represent a monomial in the new polynomial are removed from the hash table. As an example, there is a key for abc monomial in P. However, this key does not represent any monomials in P_1 after the substitution, and it has to be removed from the hash table.

7.3.2 Reverse Engineering

The proposed reverse engineering technique in Algorithm 3 (see Sect. 5.4) to identify atomic blocks requires cut enumeration. Several cut enumeration methods have been proposed to detect all n-input cuts. Some of the methods use exhaustive computation [16, 62] to detect cuts, and some others take advantage of the factor cut concept [13] to speed up the process. The efficient implementation of the cut enumeration algorithm is essential to guarantee fast reverse engineering. Such an implementation is presented in mockturtle library [80] by the function `mockturtle::cut_enumeration()`. The function traverses all nodes in the topological order and computes a node's cuts based on its fanins' cuts.

Mockturtle is a C++-17 header-only logic network library. It provides several logic network implementations (such as AIGs, MIGs, and k-LUT networks) and generic algorithms for logic synthesis and logic optimization. We have included the mockturtle library in RevSCA-2.0 and used the cut enumeration function for reverse engineering. We have also taken advantage of AIG classes in mockturtle to read AIG files.

7.4 Multiplier Generator

In the last three chapters, we depicted in theory how the integration of reverse engineering, local vanishing monomials removal, and dynamic backward rewriting into SCA verification helps in proving the correctness of structurally complex multipliers. Additionally, it is also necessary to show the efficiency of the proposed techniques by experimental evaluations. To this end, we need a wide variety of multiplier architectures with different sizes to challenge our SCA-based verifier RevSCA-2.0.

Thus, we propose the multiplier generator GenMul, which can generate very different multiplier architectures based on a wide set of algorithms. Related to GenMul is the Arithmetic Module Generator, known as AOKI [4]. However, AOKI does only support a maximum input width of 64 bits per multiplier input and is also closed-source. In contrast, the input size of GenMul is not bounded, and we have made GenMul open-source under MIT-license.[1]

In this section, we first provide an overview of GenMul, including the main data structures. Then, we describe the supported multiplier architectures that can be generated in the form of Verilog netlists using GenMul.

7.4.1 Overview and Data Structures

The main idea behind most of the multiplication algorithms is to generate and reduce partial products. Partial products are characterized by a *weight*[2] w. Therefore, the integer value of a partial product is $p2^w$, where p is a Boolean number. The partial products with the same weights can be added together to generate the reduced

[1] GenMul is available on http://www.sca-verification.org/genmul.

[2] In literature sometimes also termed significance.

set. In the C++ implementation of GENMUL, we have defined a class named `Partial` containing the two data members `weight` and `ID`, respectively. `ID` is an integer identification number assigned to a partial product automatically during its initialization. `ID` numbers are unique for each partial product and help us to implement them as wires with specific names in the final Verilog file.

The generation and reduction of partial products are performed by some computational components in each stage of a multiplier. In the first stage, AND gates are the main components. Half-adders, full-adders, and other larger adders are used in the second stage to add the partial products. Finally, multiplexers and carry propagation hardware as well as half-adders and full-adders construct the final stage of a multiplier. A parent class named `Component` has been defined in GENMUL covering all possible existing computational components. This class has two data members, `inputs` and `outputs`, which are vectors of partials and play the role of component inputs and outputs. An existing computational component, e.g., a full-adder, is implemented as a class inheriting from the parent class `Component`. All these classes contain functions to evaluate outputs based on the partial inputs and the type of the component and also generate the Verilog code of the component. Besides this, it is possible for the user to add new components to GENMUL and use them for implementing new architectures.

The availability of the `Partial` and `Component` classes in GENMUL allows to easily add further multiplier algorithms to form new multiplier architectures. For example, the Wallace tree takes advantage of a parallel addition algorithm to reduce partial products using half-adders and full-adders. Hence, the key steps in the implementation are:

1. Adding the partial products with the same weights according to the Wallace algorithm
2. Collecting the outputs of components as new partial products and hashing them based on the weights
3. Repeating the first and second steps until having a maximum of two partial products with the same weights

7.4.2 Generation of Multipliers

GENMUL already supports several architectures/algorithms for each stage of a multiplier. We are also working on extra architectures to increase the diversity of the designs even more. Table 7.2 shows the multiplier architectures of GENMUL for three stages of a multiplier. All architectures shown in black are ready to be used. The architectures with gray background are under development and will be available soon.

One of the main features of GENMUL is the generation of multipliers with arbitrary input sizes. Moreover, GENMUL is open-source on GitHub, and due to the generic data structures, new architectures/algorithms can be easily added. Finally,

Table 7.2 Multiplier architectures of GENMUL

First stage (PPG)	Second stage (PPA)	Third stage (FSA)
Unsigned simple PPG	Array	Ripple carry
Signed simple PPG	Wallace tree	Carry look-ahead
Unsigned Booth PPG	Dadda tree	Ladner–Fischer
Signed Booth PPG	Counter-based Wallace	Kogge–Stone
	Balanced delay tree	Brent–Kung
	Overturned-stairs tree	Carry-skip
		Serial prefix
		Han–Carlson
		Carry select
		Conditional sum

the web-based version of GENMUL is available. We have used the Emscripten toolchain [30] to compile Javascript from our C++ implementation of GENMUL. This allows us to configure an available multiplier of GENMUL on our webpage. Then, after pressing the "Generate" button (see Fig. A.1), the requested Verilog file can be directly downloaded.

7.5 Experimental Results

In this section, we evaluate the efficiency of REVSCA-2.0 in verifying a wide range of multiplier architectures with different sizes. First, the general details throughout the experiments, such as the machine specifications, are given. Then, we report the results of verifying clean multipliers (both structurally simple and complex), including the run-times and verification information. Subsequently, we report the results of verifying dirty optimized multipliers. The verification run-times of REVSCA-2.0 for both clean and dirty multipliers are compared with a commercial tool as well as the state-of-the-art SCA-based verification approaches.

7.5.1 General Details

The general details of verification and benchmarks throughout the experiments are listed as follows:

- All experiments have been carried out on an Intel Xeon E3-1270 v3 with 3.50 GHz and 32 GByte of main memory.
- The abbreviations for the used architectures are found in Table 2.2.

- To generate the multipliers up to 64 × 64 input sizes, we use the *Arithmetic Module Generator* [4] known as AOKI, which supports a wide range of architectures. However, AOKI cannot generate multipliers bigger than 64 × 64. Therefore, we also employ our multiplier generator GenMul to create large multipliers with 128 × 128, 256 × 256, and 512 × 512 input sizes.
- The generated multipliers are in the form of RTL Verilog code. We run Yosys [89] and abc [1] to synthesize them to an AIG representation.
- *Time-out* (T.O.) has been set to 48 h.
- *Not Supported* (N.S.) indicates that the method does not support the verification of the benchmark.

7.5.2 Clean Multipliers

We first report the results of verifying clean not-optimized multipliers with different sizes using REVSCA-2.0.

Tables 7.3 and 7.4 report the verification times for the unsigned and signed multipliers generated by AOKI, respectively. On the other hand, Tables 7.5 and 7.6 give the verification time for the unsigned and signed multipliers generated by GENMUL, respectively. The first column of the tables **Benchmark** presents the type of the multiplier based on the stage architectures. The second column **Size** denotes the size of the multiplier based on the two inputs' bit-width. The run-time (in seconds) of our proposed method is reported in detail in the third column **REVSCA-2.0** consisting of five subcolumns: *Reverse Engineering* reports the required time for extracting cuts in the AIG representation of a multiplier and then identifying atomic blocks. *Cone Detection* refers to the time needed for detecting CNCs and FFCs. *Local Van. Removal* presents the consumed time for extracting the polynomial for each CNC and FFC cone and removing vanishing monomials. *Dynamic Backw. Rewriting* reports the time for the dynamic backward rewriting phase. Finally, the overall run-time of our proposed method is presented in *Overall*. The fourth column **State-of-the-art methods** of tables reports the run-times of the state-of-the-art verification methods. This column consists of seven subcolumns: *Comm.* refers to the run-time of a commercial formal verification tool. The remaining columns show the run-times of the most recent SCA-based verification approaches.

It is evident in the tables that our proposed approach can verify all benchmarks including both unsigned and signed multipliers with different sizes. The reverse engineering time in most of the cases is small compared to the overall verification time as the algorithm to extract cuts and our proposed algorithm to identify atomic blocks are very fast. The only exceptions are $SP{\circ}CT{\circ}BK$ and $BP{\circ}CT{\circ}BK$, which require the extraction of cuts with five inputs as they contain (4:2) compressors. Therefore, the reverse engineering time increases for these benchmarks. The most time-consuming verification phase is the dynamic backward rewriting as it requires many polynomial calculations (e.g., polynomial substitution) and several possible backtracks (i.e., restoring SP_i if a sharp increase happens).

Table 7.3 Run-times of verifying unsigned AOKI multipliers (seconds)

		RevSCA-2.0					State-of-the-art methods						
Benchmark	Size	Reverse engineering	Cone detection	Local van. removal	Dynamic backw. rewriting	Overall	Comm.	[31]	[71]	[94]	[72]	[75]	[41]
$SP \circ BD \circ KS$	16×16	0.04	0.00	0.02	0.03	0.09	48.00	T.O.	T.O.	T.O.	T.O.	T.O.	0.10
$BP \circ DT \circ LF$	16×16	0.06	0.00	0.01	0.06	0.13	53.00	T.O.	T.O.	T.O.	T.O.	2.89	0.07
$SP \circ AR \circ RC$	32×32	0.12	0.00	0.00	0.17	0.29	T.O.	3.72	39.73	0.02	1.07	23.58	0.15
$SP \circ OS \circ CU$		0.21	0.01	0.01	0.44	0.66	T.O.	T.O.	T.O.	T.O.	T.O.	T.O.	0.18
$SP \circ DT \circ LF$		0.14	0.00	0.01	0.28	0.42	T.O.	T.O.	T.O.	T.O.	T.O.	61.20	0.23
$SP \circ WT \circ CL$		0.27	0.02	0.13	0.54	0.96	T.O.	T.O.	T.O.	T.O.	T.O.	T.O.	0.52
$SP \circ BD \circ KS$		0.18	0.01	0.25	0.41	0.84	T.O.	T.O.	T.O.	T.O.	T.O.	T.O.	0.29
$SP \circ CT \circ BK$		0.74	0.01	0.01	0.56	1.33	T.O.	T.O.	T.O.	T.O.	T.O.	52.18	0.16
$SP \circ AR \circ BL$		0.13	0.00	0.00	0.25	0.38	T.O.	T.O.	T.O.	T.O.	T.O.	T.O.	T.O.
$SP \circ OS \circ RL$		0.18	0.00	0.01	0.29	0.48	T.O.	T.O.	T.O.	T.O.	T.O.	T.O.	0.18
$BP \circ AR \circ RC$	32×32	0.16	0.00	0.01	0.72	0.89	T.O.	2.90	T.O.	0.02	T.O.	25.05	0.15
$BP \circ OS \circ CU$		0.22	0.01	0.01	1.17	1.41	T.O.	T.O.	T.O.	T.O.	T.O.	T.O.	0.18
$BP \circ DT \circ LF$		0.13	0.01	0.01	0.79	0.94	T.O.	T.O.	T.O.	T.O.	T.O.	T.O.	0.20
$BP \circ WT \circ CL$		0.30	0.03	0.16	0.90	1.38	T.O.	T.O.	T.O.	T.O.	T.O.	T.O.	0.50
$BP \circ BD \circ KS$		0.20	0.01	0.26	0.78	1.25	T.O.	T.O.	T.O.	T.O.	T.O.	T.O.	0.29
$BP \circ CT \circ BK$		0.45	0.01	0.02	1.03	1.50	T.O.	T.O.	T.O.	T.O.	T.O.	50.14	0.18
$BP \circ AR \circ BL$		0.20	0.00	0.01	0.76	0.97	T.O.	T.O.	T.O.	T.O.	T.O.	T.O.	T.O.
$BP \circ OS \circ RL$		0.20	0.01	0.01	0.81	1.03	T.O.	T.O.	T.O.	T.O.	T.O.	T.O.	0.21

$SP \circ AR \circ RC$	64×64	0.56	0.01	0.01	2.53	3.10	T.O.	49.91	T.O.	0.14	14.17	781.12	0.75
$SP \circ OS \circ CU$		1.18	0.02	0.02	6.67	7.88	T.O.	T.O.	T.O.	T.O.	T.O.	T.O.	0.96
$SP \circ DT \circ LF$		0.70	0.01	0.02	4.16	4.90	T.O.	T.O.	T.O.	T.O.	T.O.	2105.74	1.16
$SP \circ WT \circ CL$		2.05	0.34	2.49	10.29	15.17	T.O.	T.O.	T.O.	T.O.	T.O.	T.O.	4.24
$SP \circ BD \circ KS$		0.99	0.04	3.79	6.54	11.34	T.O.	T.O.	T.O.	T.O.	T.O.	T.O.	1.58
$SP \circ CT \circ BK$		12.94	0.05	0.06	9.82	22.88	T.O.	T.O.	T.O.	T.O.	T.O.	1726.67	0.73
$SP \circ AR \circ BL$		0.64	0.01	0.01	3.10	3.76	T.O.	T.O.	T.O.	T.O.	T.O.	T.O.	T.O.
$SP \circ OS \circ RL$		1.04	0.01	0.04	6.05	7.14	T.O.	T.O.	T.O.	T.O.	T.O.	T.O.	1.00
$BP \circ AR \circ RC$	64×64	1.33	0.02	0.03	12.59	13.97	T.O.	37.18	T.O.	0.09	T.O.	882.52	0.91
$BP \circ OS \circ CU$		1.55	0.03	0.04	21.46	23.08	T.O.	T.O.	T.O.	T.O.	T.O.	T.O.	1.09
$BP \circ DT \circ LF$		0.74	0.02	0.05	13.50	14.32	T.O.	T.O.	T.O.	T.O.	T.O.	T.O.	1.16
$BP \circ WT \circ CL$		2.37	0.38	2.73	15.45	20.91	T.O.	T.O.	T.O.	T.O.	T.O.	T.O.	5.15
$BP \circ BD \circ KS$		1.28	0.05	3.92	13.23	18.47	T.O.	T.O.	T.O.	T.O.	T.O.	T.O.	1.57
$BP \circ CT \circ BK$		6.96	0.04	0.06	20.66	27.72	T.O.	T.O.	T.O.	T.O.	T.O.	1729.33	0.99
$BP \circ AR \circ BL$		1.63	0.02	0.05	15.22	16.91	T.O.	T.O.	T.O.	T.O.	T.O.	T.O.	T.O.
$BP \circ OS \circ RL$		1.35	0.02	0.06	17.71	19.15	T.O.	T.O.	T.O.	T.O.	T.O.	T.O.	1.07

Table 7.4 Run-times of verifying signed AOKI multipliers (seconds)

		REVSCA-2.0					State-of-the-art methods						
Benchmark	Size	Reverse engineering	Cone detection	Local van. removal	Dynamic backw. rewriting	Overall	Comm.	[31]	[71]	[94]	[72]	[75]	[41]
$SP \circ BD \circ KS$	16×16	0.06	0.00	0.02	0.03	0.12	61.00	N.S.	N.S.	N.S.	N.S.	N.S.	0.09
$BP \circ DT \circ LF$	16×16	0.07	0.00	0.01	0.08	0.16	58.00	N.S.	N.S.	N.S.	N.S.	N.S.	0.06
$SP \circ AR \circ RC$	32×32	0.12	0.00	0.00	0.16	0.28	T.O.	N.S.	N.S.	N.S.	N.S.	N.S.	0.15
$SP \circ OS \circ CU$		0.20	0.00	0.01	0.45	0.66	T.O.	N.S.	N.S.	N.S.	N.S.	N.S.	0.21
$SP \circ DT \circ LF$		0.14	0.00	0.01	0.28	0.42	T.O.	N.S.	N.S.	N.S.	N.S.	N.S.	0.21
$SP \circ WT \circ CL$		0.27	0.02	0.13	0.55	0.97	T.O.	N.S.	N.S.	N.S.	N.S.	N.S.	0.50
$SP \circ BD \circ KS$		0.17	0.01	0.25	0.41	0.84	T.O.	N.S.	N.S.	N.S.	N.S.	N.S.	0.30
$SP \circ CT \circ BK$		0.74	0.01	0.01	0.57	1.33	T.O.	N.S.	N.S.	N.S.	N.S.	N.S.	0.16
$SP \circ AR \circ BL$		0.13	0.00	0.00	0.24	0.38	N.S.	N.S.	N.S.	N.S.	N.S.	T.O.	0.17
$SP \circ OS \circ RL$		0.18	0.00	0.01	0.29	0.48	T.O.	N.S.	N.S.	N.S.	N.S.	N.S.	0.24
$BP \circ AR \circ RC$	32×32	0.14	0.00	0.01	0.67	0.83	T.O.	N.S.	N.S.	N.S.	N.S.	N.S.	0.16
$BP \circ OS \circ CU$		0.21	0.01	0.01	1.12	1.35	T.O.	N.S.	N.S.	N.S.	N.S.	N.S.	0.17
$BP \circ DT \circ LF$		0.12	0.01	0.01	0.83	0.97	T.O.	N.S.	N.S.	N.S.	N.S.	N.S.	0.15
$BP \circ WT \circ CL$		0.29	0.03	0.16	0.86	0.33	T.O.	N.S.	N.S.	N.S.	N.S.	N.S.	0.49
$BP \circ BD \circ KS$		0.19	0.01	0.26	0.77	1.23	T.O.	N.S.	N.S.	N.S.	N.S.	N.S.	0.31
$BP \circ CT \circ BK$		0.37	0.01	0.01	0.94	1.33	T.O.	N.S.	N.S.	N.S.	N.S.	N.S.	0.18
$BP \circ AR \circ BL$		0.19	0.00	0.01	0.75	0.96	T.O.	N.S.	N.S.	N.S.	N.S.	N.S.	0.17
$BP \circ OS \circ RL$		0.18	0.01	0.01	0.79	0.99	T.O.	N.S.	N.S.	N.S.	N.S.	N.S.	0.21

$SP \circ AR \circ RC$	64×64	0.56	0.01	0.01	2.55	3.12	T.O.	N.S.	N.S.	N.S.	N.S.	N.S.	0.80
$SP \circ OS \circ CU$		1.18	0.02	0.02	6.72	7.94	T.O.	N.S.	N.S.	N.S.	N.S.	N.S.	0.95
$SP \circ DT \circ LF$		0.69	0.01	0.02	4.19	4.91	T.O.	N.S.	N.S.	N.S.	N.S.	N.S.	1.13
$SP \circ WT \circ CL$		2.06	0.34	2.51	10.47	15.37	T.O.	N.S.	N.S.	N.S.	N.S.	N.S.	4.24
$SP \circ BD \circ KS$		0.99	0.04	3.78	6.70	11.50	T.O.	N.S.	N.S.	N.S.	N.S.	N.S.	1.58
$SP \circ CT \circ BK$		12.92	0.05	0.06	9.94	22.97	T.O.	N.S.	N.S.	N.S.	N.S.	N.S.	0.77
$SP \circ AR \circ BL$		0.65	0.01	0.01	3.14	3.82	T.O.	N.S.	N.S.	N.S.	N.S.	N.S.	0.85
$SP \circ OS \circ RL$		1.06	0.01	0.04	6.13	7.24	T.O.	N.S.	N.S.	N.S.	N.S.	N.S.	1.02
$BP \circ AR \circ RC$	64×64	1.25	0.02	0.03	12.30	13.60	T.O.	N.S.	N.S.	N.S.	N.S.	N.S.	0.90
$BP \circ OS \circ CU$		1.51	0.03	0.04	21.31	22.89	T.O.	N.S.	N.S.	N.S.	N.S.	N.S.	1.08
$BP \circ DT \circ LF$		0.73	0.02	0.05	14.10	14.91	T.O.	N.S.	N.S.	N.S.	N.S.	N.S.	T.O.
$BP \circ WT \circ CL$		2.26	0.38	2.71	15.34	20.69	T.O.	N.S.	N.S.	N.S.	N.S.	N.S.	5.15
$BP \circ BD \circ KS$		1.24	0.05	3.92	12.81	18.01	T.O.	N.S.	N.S.	N.S.	N.S.	N.S.	1.53
$BP \circ CT \circ BK$		6.13	0.04	0.06	19.28	25.51	T.O.	N.S.	N.S.	N.S.	N.S.	N.S.	0.98
$BP \circ AR \circ BL$		1.56	0.02	0.05	15.31	16.94	T.O.	N.S.	N.S.	N.S.	N.S.	N.S.	0.97
$BP \circ OS \circ RL$		1.32	0.02	0.06	17.78	19.19	T.O.	N.S.	N.S.	N.S.	N.S.	N.S.	1.03

Table 7.5 Run-times of verifying unsigned GENMUL multipliers (seconds)

		REVSCA-2.0					State-of-the-art methods						
Benchmark	Size	Reverse engineering	Cone detection	Local van. removal	Dynamic backw. rewriting	Overall	Comm.	[31]	[71]	[94]	[72]	[75]	[41]
$SP \circ AR \circ RC$	128×128	2.72	0.05	0.02	78.42	81.22	T.O.	T.O.	T.O.	T.O.	T.O.	T.O.	5.56
$SP \circ WT \circ BK$		7.54	0.06	0.04	134.58	142.22	T.O.	T.O.	T.O.	T.O.	T.O.	T.O.	7.13
$SP \circ DT \circ LF$		3.71	0.07	0.08	149.74	153.60	T.O.	T.O.	T.O.	T.O.	T.O.	T.O.	7.22
$SP \circ AR \circ RC$	256×256	20.01	0.28	0.11	2469.11	2489.51	T.O.	T.O.	T.O.	T.O.	T.O.	T.O.	55.58
$SP \circ WT \circ BK$		72.15	0.32	0.16	3700.59	3773.21	T.O.	T.O.	T.O.	T.O.	T.O.	T.O.	67.84
$SP \circ DT \circ LF$		27.84	0.35	0.32	5593.94	5622.45	T.O.	T.O.	T.O.	T.O.	T.O.	T.O.	66.96
$SP \circ AR \circ RC$	512×512	422.32	1.38	0.42	48,338.40	48,762.60	T.O.	T.O.	T.O.	T.O.	T.O.	T.O.	588.28
$SP \circ WT \circ BK$		960.27	1.60	0.59	66,530.90	67,493.30	T.O.	T.O.	T.O.	T.O.	T.O.	T.O.	640.90
$SP \circ DT \circ LF$		480.85	1.73	1.29	113,774.00	114,257.87	T.O.	T.O.	T.O.	T.O.	T.O.	T.O.	790.12

Table 7.6 Run-times of verifying signed GENMUL multipliers (seconds)

		REVSCA-2.0					State-of-the-art methods						
Benchmark	Size	Reverse engineering	Cone detection	Local van. removal	Dynamic backw. rewriting	Overall	Comm.	[31]	[71]	[94]	[72]	[75]	[41]
$SP \circ AR \circ RC$	128×128	3.51	0.06	0.03	102.59	106.19	T.O.	N.S.	N.S.	N.S.	N.S.	N.S.	5.39
$SP \circ WT \circ BK$		8.39	0.08	0.04	180.83	189.34	T.O.	N.S.	N.S.	N.S.	N.S.	N.S.	10.32
$SP \circ DT \circ LF$		4.40	0.08	0.08	217.65	222.22	T.O.	N.S.	N.S.	N.S.	N.S.	N.S.	10.03
$SP \circ AR \circ RC$	256×256	26.78	0.29	0.10	2408.70	2435.87	T.O.	N.S.	N.S.	N.S.	N.S.	N.S.	46.03
$SP \circ WT \circ BK$		76.84	0.32	0.15	3676.48	3753.80	T.O.	N.S.	N.S.	N.S.	N.S.	N.S.	87.40
$SP \circ DT \circ LF$		34.83	0.36	0.32	6059.79	6095.31	T.O.	N.S.	N.S.	N.S.	N.S.	N.S.	100.44
$SP \circ AR \circ RC$	512×512	672.67	1.44	0.41	46432.70	47,107.30	T.O.	N.S.	N.S.	N.S.	N.S.	N.S.	427.58
$SP \circ WT \circ BK$		1184.83	1.65	0.57	67,076.30	68,263.40	T.O.	N.S.	N.S.	N.S.	N.S.	N.S.	879.82
$SP \circ DT \circ LF$		740.61	1.76	1.30	114,750.00	115,493.00	T.O.	N.S.	N.S.	N.S.	N.S.	N.S.	841.99

On the other hand, the commercial tool only verifies multipliers up to 16×16, and it times out for the bigger benchmarks. The proposed SCA-based verification methods of [31, 71, 72, 94] either cannot verify any benchmarks or only work on the structurally simple multipliers, i.e., $SP{\circ}AR{\circ}RC$ and $BP{\circ}AR{\circ}RC$. The main reason is that these methods do not provide any solution to remove vanishing monomials early in the calculations to avoid explosion during global backward rewriting. The proposed method in [75] can verify some of the structurally complex multipliers as the authors presented a heuristic to detect and remove vanishing monomials. However, it is not robust as can be seen in column [75] and fails for most of the benchmarks. It is also drastically slower than our proposed method. In addition, the aforementioned SCA-based methods do not support the verification of signed multipliers (see Tables 7.4 and 7.6).

The proposed method in [41] reports very good results for the verification of unsigned and signed multipliers if the FSA can be detected. However, it fails to verify five AOKI benchmarks including four unsigned multipliers (i.e., $SP{\circ}AR{\circ}BL$ and $BP{\circ}AR{\circ}BL$ with 32×32 and 64×64 input sizes) and one signed multiplier (i.e., $BP{\circ}DT{\circ}LF$) with 64×64 input size. Hence, the method is faster than REVSCA-2.0 for clean multipliers, but it supports the verification of fewer architectures.

Tables 7.7 and 7.8 present the verification information reported by REVSCA-2.0 for unsigned and signed AOKI multipliers, respectively. On the other hand, Tables 7.9 and 7.10 give the verification information reported by REVSCA-2.0 for unsigned and signed GENMUL multipliers, respectively. The first and second columns of the tables show the type and the size of the multiplier, respectively. The third column **#Nodes** reports the number of nodes in the AIG representation of the multiplier. The number of identified atomic blocks is presented in the fourth column **#Atomic**. The fifth column **#Van.** gives the total number of canceled vanishing monomials in the local vanishing removal phase. Finally, the sixth column **MaxPoly** reports the maximum size of SP_i during dynamic backward rewriting by counting the number of monomials.

The results in the tables confirm that REVSCA-2.0 can verify multipliers with more than 3M AIG nodes, e.g., the signed $SP{\circ}WT{\circ}BK$ multiplier in Table 7.10 contains 3,161,854 nodes. The number of detected atomic blocks varies based on the size of the multiplier and its architecture. For example, the multipliers with radix-4 Booth encoding in the PPG stage have fewer atomic blocks compared to those that use the simple PPG. The reason is that the number of generated partial products is smaller in the case of Booth encoding; thus, fewer atomic blocks are required to reduce these partial products. The total number of canceled vanishing monomials also varies based on the architecture. No vanishing monomial is generated during the verification of structurally simple multipliers, e.g., $SP{\circ}AR{\circ}RC$ and $BP{\circ}AR{\circ}RC$; therefore, the number of canceled vanishing monomials is zero. On the other hand, during the verification of 64×64 $BP{\circ}WT{\circ}CL$ and $BP{\circ}BD{\circ}KS$, which are structurally complex multipliers, approximately $280K$ and $616K$ vanishing monomials are canceled, respectively.

Table 7.7 Verification info of unsigned AOKI multipliers

Benchmark	Size	#Nodes	#Atomic	#Van.	MaxPoly
$SP \circ BD \circ KS$	16×16	3144	281	7716	444
$BP \circ DT \circ LF$	16×16	2571	166	803	1186
$SP \circ AR \circ RC$	32×32	11,776	992	0	1088
$SP \circ OS \circ CU$		13191	1192	0	2571
$SP \circ DT \circ LF$		12,014	997	1438	1649
$SP \circ WT \circ CL$		16,276	1114	27,612	1200
$SP \circ BD \circ KS$		12,780	1109	69,804	1754
$SP \circ CT \circ BK$		10,347	1071	463	1456
$SP \circ AR \circ BL$		11,814	1015	48	1377
$SP \circ OS \circ RL$		12,543	1151	497	1692
$BP \circ AR \circ RC$	32×32	9975	724	0	4931
$BP \circ OS \circ CU$		10,661	756	0	5572
$BP \circ DT \circ LF$		9619	590	1438	4930
$BP \circ WT \circ CL$		14,211	694	30,744	3187
$BP \circ BD \circ KS$		10,414	700	70,299	3190
$BP \circ CT \circ BK$		8751	671	463	4930
$BP \circ AR \circ BL$		10,016	755	98	4931
$BP \circ OS \circ RL$		10,007	715	497	4931
$SP \circ AR \circ RC$	64×64	48,128	4032	0	4224
$SP \circ OS \circ CU$		51,766	4502	0	9825
$SP \circ DT \circ LF$		48,808	4038	2249	7256
$SP \circ WT \circ CL$		68,875	4365	266,684	4461
$SP \circ BD \circ KS$		50,756	4313	613,454	5607
$SP \circ CT \circ BK$		41,466	4253	808	6080
$SP \circ AR \circ BL$		48,212	4079	96	4589
$SP \circ OS \circ RL$		49,989	4,412	929	7929
$BP \circ AR \circ RC$	64×64	38,439	2732	0	20,099
$BP \circ OS \circ CU$		39,798	2667	0	25,803
$BP \circ DT \circ LF$		36,867	2205	2249	20,098
$BP \circ WT \circ CL$		57,684	2513	280,604	12,530
$BP \circ BD \circ KS$		39,053	2514	616,166	12,532
$BP \circ CT \circ BK$		33,172	2433	815	20,101
$BP \circ AR \circ BL$		38,555	2799	170	20,099
$BP \circ OS \circ RL$		38,050	2577	909	20,100

7.5.3 Dirty Optimized Multipliers

We now report the results of verifying dirty optimized multipliers with different sizes using REVSCA-2.0.

Tables 7.11 and 7.12 report the verification results for the unsigned multipliers optimized by *resyn3* and *dc2* commands in abc. These commands are well-known

Table 7.8 Verification info of signed AOKI multipliers

Benchmark	Size	#Nodes	#Atomic	#Van.	MaxPoly
$SP \circ BD \circ KS$	16×16	3146	281	7716	445
$BP \circ DT \circ LF$	16×16	2324	144	803	1188
$SP \circ AR \circ RC$	32×32	11,778	992	0	1089
$SP \circ OS \circ CU$		13,193	1192	0	2571
$SP \circ DT \circ LF$		12,015	997	1438	1649
$SP \circ WT \circ CL$		16,278	1114	27,612	1199
$SP \circ BD \circ KS$		12,782	1109	69,804	1683
$SP \circ CT \circ BK$		10,347	1071	463	1456
$SP \circ AR \circ BL$		11,816	1015	48	1377
$SP \circ OS \circ RL$		12,545	1151	497	1692
$BP \circ AR \circ RC$	64×64	9399	664	0	4932
$BP \circ OS \circ CU$		10,135	707	0	5574
$BP \circ DT \circ LF$		9099	543	1438	4932
$BP \circ WT \circ CL$		13,687	646	30,744	3189
$BP \circ BD \circ KS$		9886	649	70,299	3191
$BP \circ CT \circ BK$		8371	622	463	4932
$BP \circ AR \circ BL$		9460	698	68	4932
$BP \circ OS \circ RL$		9487	666	497	4932
$SP \circ AR \circ RC$	64×64	48,130	4032	0	4225
$SP \circ OS \circ CU$		51,768	4502	0	9825
$SP \circ DT \circ LF$		48,809	4038	2249	7256
$SP \circ WT \circ CL$		68,877	4365	266,684	4460
$SP \circ BD \circ KS$		50,758	4313	613,454	5742
$SP \circ CT \circ BK$		41,466	4253	808	6080
$SP \circ AR \circ BL$		48,214	4079	96	4589
$SP \circ OS \circ RL$		49,991	4412	929	7929
$BP \circ AR \circ RC$	64×64	37,255	2608	0	20,100
$BP \circ OS \circ CU$		38,739	2566	0	25,805
$BP \circ DT \circ LF$		35,806	2110	2249	20,100
$BP \circ WT \circ CL$		56,614	2411	280,604	12,531
$BP \circ BD \circ KS$		37,981	2413	616,166	12,533
$BP \circ CT \circ BK$		32,412	2335	815	20,100
$BP \circ AR \circ BL$		37,391	2678	140	20,100
$BP \circ OS \circ RL$		36,991	2476	909	20,100

combinational AIG optimizations, which are used to reduce the overall circuit's delay. The first column **Benchmark** of the tables lists the architecture of the multiplier based on its stages. The second column **Size** shows the size of the multiplier based on the two inputs' bit-width.

The verification data of REVSCA-2.0 is reported in the third column **Verification data** that consists of three subcolumns: *#Nodes* shows the number of AIG nodes

Table 7.9 Verification info of unsigned GENMUL multipliers

Benchmark	Size	#Nodes	#Atomic	#Van.	MaxPoly
$SP \circ AR \circ RC$	128×128	162,304	16,256	0	16,640
$SP \circ WT \circ BK$		166,938	17,366	1623	22,406
$SP \circ DT \circ LF$		164,572	16,263	3642	29,811
$SP \circ AR \circ RC$	256×256	652,288	65,280	0	66,048
$SP \circ WT \circ BK$		663,505	67,974	3376	81,460
$SP \circ DT \circ LF$		657,622	65,288	5800	120,738
$SP \circ AR \circ RC$	512×512	2,615,296	261,632	0	261,632
$SP \circ WT \circ BK$		2,641,643	267,985	6851	296,667
$SP \circ DT \circ LF$		2,627,536	261,641	9718	485,741

Table 7.10 Verification info of signed GENMUL multipliers

Benchmark	Size	#Nodes	#Atomic	#Van.	MaxPoly
$SP \circ AR \circ RC$	128×128	194,562	16,256	0	16,641
$SP \circ WT \circ BK$		198,697	17,366	1623	22,449
$SP \circ DT \circ LF$		196,322	16,263	3642	29,811
$SP \circ AR \circ RC$	256×256	782,338	65,280	0	66,049
$SP \circ WT \circ BK$		792,538	67,974	3376	81,552
$SP \circ DT \circ LF$		786,653	65,288	5800	120,738
$SP \circ AR \circ RC$	512×512	3,137,538	261,632	0	263,169
$SP \circ WT \circ BK$		3,161,854	267,986	6851	296,728
$SP \circ DT \circ LF$		3,147,734	261,641	9718	485,741

of the multiplier, *#Van.* gives the total number of removed vanishing monomials, and *MaxPoly* shows the maximum size of the intermediate polynomial SP_i during backward rewriting by counting the number of monomials. The fourth column **Run-times** of Tables 7.11 and 7.12 reports the overall run-time of our SCA-based verifier REVSCA-2.0 and the other state-of-the-art formal verification methods.

As can be seen in Table 7.11, REVSCA-2.0 only produces a time-out for the multipliers with $SP \circ WT \circ CL$ and $BP \circ WT \circ CL$ architectures when they are optimized using *resyn3*, and it successfully proves the correctness of the other optimized benchmarks. When it comes to the *dc2* optimization in Table 7.12, REVSCA-2.0 produces time-out for the $SP \circ BD \circ KS$ and $BP \circ BD \circ KS$ benchmarks in addition to the two mentioned architectures, and it successfully verifies the other optimized multipliers. Consequently, REVSCA-2.0 can verify a wide range of optimized multipliers with different architectures. Statistically, from the total 74 optimized benchmarks in Tables 7.11 and 7.12, REVSCA-2.0 successfully verifies 61 benchmarks and only times-out for 13 benchmarks, i.e., it has the 85% success rate, which is by far the best among the existing formal verification methods.

On the other hand, the commercial tool can only verify the two optimized multipliers with 16×16 input size, and it fails for the bigger benchmarks. Despite the good results of [41] for the clean multipliers, this method only verifies one

Table 7.11 Results of verifying unsigned multipliers optimized by *resyn3*

Benchmark	Size	Verification data			Run-times (seconds)							
		#Nodes	#Van.	MaxPoly	RevSCA-2.0	Comm.	[31]	[71]	[94]	[72]	[75]	[41]
$SP \circ BD \circ KS$	16×16	2953	9742	488	0.09	54.00	T.O.	T.O.	T.O.	T.O.	T.O.	T.O.
$BP \circ DT \circ LF$	16×16	2418	1463	1185	0.09	57.00	T.O.	T.O.	T.O.	T.O.	2.89	T.O.
$SP \circ AR \circ RC$	32×32	10,525	0	1088	0.27	T.O.	T.O.	T.O.	T.O.	T.O.	T.O.	0.21
$SP \circ OS \circ CU$		12,404	0	1951	0.55	T.O.	T.O.	T.O.	T.O.	T.O.	T.O.	T.O.
$SP \circ DT \circ LF$		11,067	2323	1706	0.37	T.O.	T.O.	T.O.	T.O.	T.O.	T.O.	T.O.
$SP \circ WT \circ CL$		15,173	–	–	T.O.	T.O.	T.O.	T.O.	T.O.	T.O.	T.O.	T.O.
$SP \circ BD \circ KS$		12,021	118,332	1204	1.00	T.O.	T.O.	T.O.	T.O.	T.O.	T.O.	T.O.
$SP \circ CT \circ BK$		10,297	579	1584	1.63	T.O.	T.O.	T.O.	T.O.	T.O.	T.O.	T.O.
$SP \circ AR \circ BL$		11,586	65	1616	0.40	T.O.	T.O.	T.O.	T.O.	T.O.	T.O.	T.O.
$SP \circ OS \circ RL$		12,036	426	1959	0.51	T.O.	T.O.	T.O.	T.O.	T.O.	T.O.	T.O.
$BP \circ AR \circ RC$	32×32	8931	0	4929	0.97	T.O.	T.O.	T.O.	T.O.	T.O.	T.O.	T.O.
$BP \circ OS \circ CU$		9806	0	4932	1.17	T.O.	T.O.	T.O.	T.O.	T.O.	T.O.	T.O.
$BP \circ DT \circ LF$		8921	1948	4929	0.94	T.O.	T.O.	T.O.	T.O.	T.O.	T.O.	T.O.
$BP \circ WT \circ CL$		13,144	–	–	T.O.	T.O.	T.O.	T.O.	T.O.	T.O.	T.O.	T.O.
$BP \circ BD \circ KS$		9800	129,608	3194	1.65	T.O.	T.O.	T.O.	T.O.	T.O.	T.O.	T.O.
$BP \circ CT \circ BK$		8632	580	4929	1.62	T.O.	T.O.	T.O.	T.O.	T.O.	T.O.	T.O.
$BP \circ AR \circ BL$		9526	145	5066	1.02	T.O.	T.O.	T.O.	T.O.	T.O.	T.O.	T.O.
$BP \circ OS \circ RL$		9443	447	4932	1.00	T.O.	T.O.	T.O.	T.O.	T.O.	T.O.	T.O.

$SP \circ AR \circ RC$	64×64	42,365	0	4224	3.14	T.O.	T.O.	T.O.	T.O.	T.O.	T.O.	1.02
$SP \circ OS \circ CU$		48,694	0	7758	6.48	T.O.	T.O.	T.O.	T.O.	T.O.	T.O.	T.O.
$SP \circ DT \circ LF$		45,115	3038	6908	5.06	T.O.	T.O.	T.O.	T.O.	T.O.	T.O.	T.O.
$SP \circ WT \circ CL$		63,962	–	–	T.O.	T.O.	T.O.	T.O.	T.O.	T.O.	T.O.	T.O.
$SP \circ BD \circ KS$		47,746	877,004	7961	13.52	T.O.	T.O.	T.O.	T.O.	T.O.	T.O.	T.O.
$SP \circ CT \circ BK$		41,349	957	6761	30.57	T.O.	T.O.	T.O.	T.O.	T.O.	T.O.	T.O.
$SP \circ AR \circ BL$		47,300	131	4986	4.46	T.O.	T.O.	T.O.	T.O.	T.O.	T.O.	T.O.
$SP \circ OS \circ RL$		47,561	893	8400	6.77	T.O.	T.O.	T.O.	T.O.	T.O.	T.O.	T.O.
$BP \circ AR \circ RC$	64×64	34,317	0	20,097	15.66	T.O.	T.O.	T.O.	T.O.	T.O.	T.O.	T.O.
$BP \circ OS \circ CU$		37,156	0	20,100	17.80	T.O.	T.O.	T.O.	T.O.	T.O.	T.O.	T.O.
$BP \circ DT \circ LF$		33,658	3255	27,999	19.20	T.O.	T.O.	T.O.	T.O.	T.O.	T.O.	T.O.
$BP \circ WT \circ CL$		53,478	–	–	T.O.	T.O.	T.O.	T.O.	T.O.	T.O.	T.O.	T.O.
$BP \circ BD \circ KS$		36,469	802,110	13,342	21.32	T.O.	T.O.	T.O.	T.O.	T.O.	T.O.	T.O.
$BP \circ CT \circ BK$		32,931	952	20,392	32.14	T.O.	T.O.	T.O.	T.O.	T.O.	T.O.	T.O.
$BP \circ AR \circ BL$		36,925	244	20,594	16.52	T.O.	T.O.	T.O.	T.O.	T.O.	T.O.	T.O.
$BP \circ OS \circ RL$		36,326	882	20,097	18.51	T.O.	T.O.	T.O.	T.O.	T.O.	T.O.	T.O.
$SP \circ AR \circ RC$	128×128	162,298	0	16,640	75.45	T.O.	T.O.	T.O.	T.O.	T.O.	T.O.	5.53
$SP \circ WT \circ BK$		154,146	1808	22,347	127.73	T.O.	T.O.	T.O.	T.O.	T.O.	T.O.	T.O.
$SP \circ DT \circ LF$		150,711	4900	28,721	146.87	T.O.	T.O.	T.O.	T.O.	T.O.	T.O.	T.O.

Table 7.12 Results of verifying unsigned multipliers optimized by *dc2*

Benchmark	Size	Verification data			Run-times (seconds)							
		#Nodes	#Van.	MaxPoly	RevSCA-2.0	Comm.	[31]	[71]	[94]	[72]	[75]	[41]
$SP \circ BD \circ KS$	16×16	2274	–	–	T.O.	54.00	T.O.	T.O.	T.O.	T.O.	T.O.	T.O.
$BP \circ DT \circ LF$	16×16	2003	54	906	0.07	57.00	T.O.	T.O.	T.O.	T.O.	2.89	T.O.
$SP \circ AR \circ RC$	32×32	8862	0	1089	0.27	T.O.	T.O.	T.O.	T.O.	T.O.	T.O.	0.18
$SP \circ OS \circ CU$		9751	0	2489	0.67	T.O.	T.O.	T.O.	T.O.	T.O.	T.O.	T.O.
$SP \circ DT \circ LF$		8902	1227	1521	0.40	T.O.	T.O.	T.O.	T.O.	T.O.	T.O.	T.O.
$SP \circ WT \circ CL$		10,696	–	–	T.O.	T.O.	T.O.	T.O.	T.O.	T.O.	T.O.	T.O.
$SP \circ BD \circ KS$		9294	–	–	T.O.	T.O.	T.O.	T.O.	T.O.	T.O.	T.O.	T.O.
$SP \circ CT \circ BK$		8194	66	1274	0.54	T.O.	T.O.	T.O.	T.O.	T.O.	T.O.	T.O.
$SP \circ AR \circ BL$		8872	175	1284	0.32	T.O.	T.O.	T.O.	T.O.	T.O.	T.O.	T.O.
$SP \circ OS \circ RL$		9285	241	1874	0.76	T.O.	T.O.	T.O.	T.O.	T.O.	T.O.	T.O.
$BP \circ AR \circ RC$	32×32	8086	0	4427	0.88	T.O.	T.O.	T.O.	T.O.	T.O.	T.O.	T.O.
$BP \circ OS \circ CU$		8455	0	5191	1.60	T.O.	T.O.	T.O.	T.O.	T.O.	T.O.	T.O.
$BP \circ DT \circ LF$		7669	1285	4032	1.70	T.O.	T.O.	T.O.	T.O.	T.O.	T.O.	T.O.
$BP \circ WT \circ CL$		9501	–	–	T.O.	T.O.	T.O.	T.O.	T.O.	T.O.	T.O.	T.O.
$BP \circ BD \circ KS$		8078	–	–	T.O.	T.O.	T.O.	T.O.	T.O.	T.O.	T.O.	T.O.
$BP \circ CT \circ BK$		7516	119	3864	1.19	T.O.	T.O.	T.O.	T.O.	T.O.	T.O.	T.O.
$BP \circ AR \circ BL$		8082	135	3850	0.90	T.O.	T.O.	T.O.	T.O.	T.O.	T.O.	T.O.
$BP \circ OS \circ RL$		7972	1720	4067	1.21	T.O.	T.O.	T.O.	T.O.	T.O.	T.O.	T.O.

$SP \circ AR \circ RC$	64×64	36,158	0	4225	3.12	T.O.	T.O.	T.O.	T.O.	T.O.	T.O.	0.82
$SP \circ OS \circ CU$		38,487	0	11,642	13.36	T.O.	T.O.	T.O.	T.O.	T.O.	T.O.	T.O.
$SP \circ DT \circ LF$		36,365	5618	8734	13.42	T.O.	T.O.	T.O.	T.O.	T.O.	T.O.	T.O.
$SP \circ WT \circ CL$		44,643	–	–	T.O.	T.O.	T.O.	T.O.	T.O.	T.O.	T.O.	T.O.
$SP \circ BD \circ KS$		37,655	–	–	T.O.	T.O.	T.O.	T.O.	T.O.	T.O.	T.O.	T.O.
$SP \circ CT \circ BK$		33,046	1133	5401	8.26	T.O.	T.O.	T.O.	T.O.	T.O.	T.O.	T.O.
$SP \circ AR \circ BL$		36,212	433	5559	4.70	T.O.	T.O.	T.O.	T.O.	T.O.	T.O.	T.O.
$SP \circ OS \circ RL$		37,311	7226	6419	5.74	T.O.	T.O.	T.O.	T.O.	T.O.	T.O.	T.O.
$BP \circ AR \circ RC$	64×64	31,312	0	15,882	14.37	T.O.	T.O.	T.O.	T.O.	T.O.	T.O.	T.O.
$BP \circ OS \circ CU$		31,925	0	26,200	42.91	T.O.	T.O.	T.O.	T.O.	T.O.	T.O.	T.O.
$BP \circ DT \circ LF$		29,855	29,305	17,748	21.52	T.O.	T.O.	T.O.	T.O.	T.O.	T.O.	T.O.
$BP \circ WT \circ CL$		38,539	–	–	T.O.	T.O.	T.O.	T.O.	T.O.	T.O.	T.O.	T.O.
$BP \circ BD \circ KS$		31,226	–	–	T.O.	T.O.	T.O.	T.O.	T.O.	T.O.	T.O.	T.O.
$BP \circ CT \circ BK$		29,141	1450	18,407	18.78	T.O.	T.O.	T.O.	T.O.	T.O.	T.O.	T.O.
$BP \circ AR \circ BL$		31,371	982	15,882	17.21	T.O.	T.O.	T.O.	T.O.	T.O.	T.O.	T.O.
$BP \circ OS \circ RL$		30,722	16,921	21,088	26.29	T.O.	T.O.	T.O.	T.O.	T.O.	T.O.	T.O.
$SP \circ AR \circ RC$	128×128	146,044	0	16,642	76.92	T.O.	T.O.	T.O.	T.O.	T.O.	T.O.	4.36
$SP \circ WT \circ BK$		146,655	43,086	57,708	1617.74	T.O.	T.O.	T.O.	T.O.	T.O.	T.O.	T.O.
$SP \circ DT \circ LF$		149,353	4160	26,911	269.58	T.O.	T.O.	T.O.	T.O.	T.O.	T.O.	T.O.

Table 7.13 Results of verifying industrial multipliers

Source	Size	#Nodes	Run-times (s)							
			RevSCA-2.0	Comm.	[31]	[71]	[94]	[72]	[75]	[41]
Synopsys DesignWare Library (pparch*)	16 × 16	2432	0.25	40.00	T.O.	T.O.	T.O.	T.O.	T.O.	T.O.
	32 × 32	7240	1.79	T.O.	T.O.	T.O.	T.O.	T.O.	T.O.	T.O.
	48 × 48	16,086	13.13	T.O.	T.O.	T.O.	T.O.	T.O.	T.O.	T.O.
	64 × 64	27,658	41.24	T.O.	T.O.	T.O.	T.O.	T.O.	T.O.	T.O.
	96 × 96	61,180	366.082	T.O.	T.O.	T.O.	T.O.	T.O.	T.O.	T.O.
	128 × 128	106,949	1468.01	T.O.	T.O.	T.O.	T.O.	T.O.	T.O.	T.O.
	160 × 160	166,492	3813.43	T.O.	T.O.	T.O.	T.O.	T.O.	T.O.	T.O.
	192 × 192	238,920	8393.28	T.O.	T.O.	T.O.	T.O.	T.O.	T.O.	T.O.
	256 × 256	422,077	26,117.7	T.O.	T.O.	T.O.	T.O.	T.O.	T.O.	T.O.
EPFL mul.	64 × 64	27,190	52.07	T.O.	T.O.	T.O.	T.O.	T.O.	T.O.	T.O.

* Delay-optimized flexible Booth Wallace after technology mapping

optimized architecture ($SP \circ AR \circ RC$), and it fails for all the other optimized multipliers in Tables 7.11 and 7.12. Hence, the method is not robust against design alterations/optimizations as they usually destroy the clean boundaries between the multiplier stages. The other state-of-the-art formal verification methods fail for all optimized multipliers in the tables.

In the last experiment, we consider industrial benchmarks from the Synopsys DesignWare Library. They have been optimized for the delay. The gate-level Verilog description of these multipliers is generated by mapping the multiplier IP to a standard cell library consisting of up to 3-input logical gates using Synopsys Design Compiler. Then, the Verilog description is converted into AIG using abc. The results can be found in Table 7.13. In addition, the table also includes the highly optimized multiplier from the EPFL combinational benchmark suite [3]. As can be seen, RevSCA-2.0 is able to prove the correctness of all these multipliers, while the commercial tool can only verify the smallest instance and all the other SCA-based methods fail.

Formal verification of industrial multipliers is one of the most notable achievements of RevSCA-2.0 and illustrates its possible applications in real-world verification challenges. It is the first step toward the formal verification of highly optimized and technology-mapped arithmetic circuits.

7.6 Conclusion

In this chapter, first, the top-level overview of our SCA-based verifier RevSCA-2.0 was given. We introduced an algorithm that describes how RevSCA-2.0 works from the moment it receives the multiplier until it proves or disproves its correctness. RevSCA-2.0 takes advantage of the three proposed techniques in the previous chapters to extend the SCA-based verification and overcome the verification challenges.

Second, the details of the REVSCA-2.0 implementation including the polynomial data structure were explained. Since polynomial computations (e.g., substitution) are the most time-consuming operations during the SCA-based verification, an efficient polynomial data structure is essential to speed up the whole process. Thus, we introduced our polynomial data structure based on hash tables. In addition, we also employed the mockturtle library in the implementation of the reverse engineering technique in REVSCA-2.0.

Third, we introduced our multiplier generator GENMUL to challenge SCA-based verification methods, including REVSCA-2.0. GENMUL supports the generation of a wide range of multipliers with arbitrary sizes. Despite other multiplier generators, GENMUL is open-source and does not have any constraints on the input size of multipliers. Therefore, we can use it to generate very large benchmarks.

Finally, we evaluated the efficiency of REVSCA-2.0 in verifying a wide variety of multiplier architectures with different sizes. REVSCA-2.0 can prove the correctness of clean not-optimized multipliers generated by AOKI or GENMUL. Moreover, it can also verify dirty multipliers optimized by *resyn3* and *dc2* AIG optimizations in abc. The experimental results clearly show that REVSCA-2.0 is the most successful and robust method when it comes to the verification of optimized architectures. REVSCA-2.0 reports very good results for the industrial multipliers as well. It proves the correctness of benchmarks from the Synopsys DesignWare Library and the EPFL combinational benchmark suite, while the other state-of-the-art methods fail to verify even the smallest benchmarks.

Chapter 8
Debugging

If verification proves that a digital circuit is buggy, then localizing and fixing bugs become the major subsequent tasks. Debugging is usually more challenging than verification since it requires finding the exact location of bugs. Especially, structurally complex multipliers are among the most difficult arithmetic circuits to debug. Most of the existing automated methods fail for these circuits; thus, this task is performed manually, which is typically very time-consuming.

In this chapter, we propose a complete debugging flow based on the combination of SCA and SAT. Complete means that our method targets the complete loop until the arithmetic circuit is guaranteed to fulfill its specification. For this, our approach consists of three phases: verification, localization, and fixing. In the experimental evaluation, we demonstrate the applicability of our approach to a wide variety of structurally complex multipliers. The proposed approach in this chapter has been published in [49].

8.1 Introduction

Formal verification methods help us to prove or disprove the correctness of a circuit. Moreover, if the circuit is buggy, these methods usually provide us with a counter-example, e.g., the remainder of backward rewriting in SCA-based verification or the input values that activate bug in SAT-based verification. A debugging method takes advantage of these counter-examples to localize and fix bugs in the design. The debugging task always needs more time and resources compared to the verification task since it has to find the exact location of bugs. Therefore, proposing an efficient debugging method is vital to speed up the task of localizing and fixing bugs.

Despite the progress in the verification of arithmetic circuits, particularly multipliers, there is a limited number of works on debugging these circuits. Authors of [79] proposed a general debugging method for digital circuits based on inserting

A. Mahzoon et al., *Formal Verification of Structurally Complex Multipliers*,
https://doi.org/10.1007/978-3-031-24571-8_8

multiplexers after each gate. The multiplexers can rewrite the values of gate outputs and help with detecting the possible faulty gates. The CNF of the new circuit after the multiplexer insertion is created, and the bug candidates are localized by SAT-solving.

Recently, researchers have proposed debugging techniques based on SCA. These techniques use the remainder of SCA-based verification to localize and fix bugs in arithmetic circuits. In [34], the authors proposed a forward rewriting in addition to backward rewriting in order to detect and fix bugs. The method calculates the difference between intermediate polynomials obtained from forward and backward rewriting at each cut. This difference gives us clues about the location of bugs. The debugging approach in [33] generates a set of tests from the remainder, which activate the bug. Then, it uses the remainder patterns to detect bugs and make the correction. The proposed method in [74] adds correction hardware to a buggy design. Therefore, the structure of the buggy circuit is preserved, and the correction hardware is added as extra gates to the output of the buggy design. Unfortunately, these debugging methods have serious limitations. They only work for a limited number of architectures. Moreover, they only detect and fix bugs that are located close to the primary inputs. As a result, their application is very restricted.

In this chapter, we come up with a complete debugging flow based on the combination of SCA and SAT. Our flow consists of three main phases:

1. **Verification:** Two verification engines based on SCA and SAT are used in parallel to verify both correct and buggy designs as fast as possible.
2. **Localization:** SCA- and SAT-based methods are used to localize the bug candidates, recursively.
3. **Fixing:** SCA- and SAT-based verifications are used in parallel to check the candidates and fix the bug.

The experimental results confirm that our proposed method can successfully debug a wide variety of structurally complex multipliers when the bug occurs in different design stages.

8.2 Fault Model

In this chapter, we consider *gate misplacement* as our fault model. This well-known fault model changes the functionality of the design by a wrong gate [33, 34, 86]. Figure 8.1a shows a 2-bit bug-free adder circuit. The gate g_4 of the circuit, which is an AND gate, is replaced with an OR gate in Fig. 8.1b. Our fault model is applicable here since one gate has been misplaced with another gate.

Such faults are likely to occur, for example, when a synthesis tool makes a mistake during the optimization of the circuit. Another prominent example of introducing such kind of faults is a bug in a multiplier generator tool, which is used to create a dedicated multiplier architecture (under given constraints).

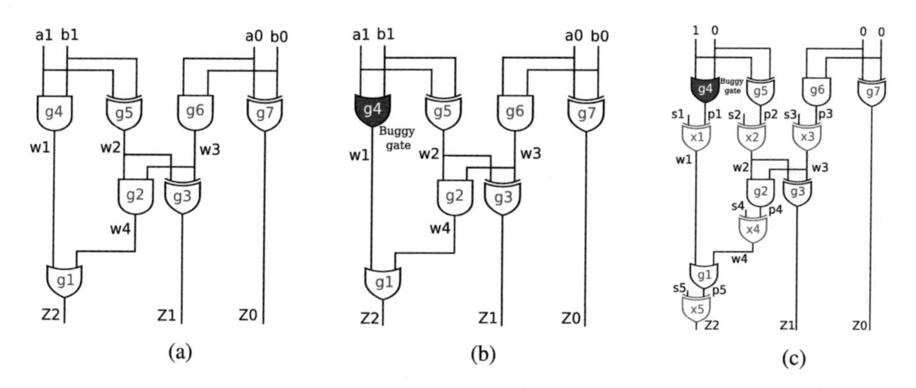

Fig. 8.1 2-bit adder circuit. (**a**) Bug-free. (**b**) Buggy at gate g_4. (**c**) After XOR insertion

8.3 Limitations of SCA-Based Debugging

In recent years, several SCA-based verification methods have been proposed to verify arithmetic circuits, particularly multipliers. We have also introduced our SCA-based verifier REVSCA-2.0 in the previous chapters. The success of SCA-based verification methods motivated researchers to extend these methods to debug arithmetic circuits. However, SCA-based approaches suffer from two major limitations when they are employed for debugging. In the next two sections, we discuss these limitations.

8.3.1 Vanishing Monomials in Remainder

In a bug-free arithmetic circuit, vanishing monomials are generated during backward rewriting, and they are reduced to zero after several substitution steps. We introduced our vanishing monomials removal technique in Chap. 4 to remove vanishing monomials locally before global backward rewriting. However, in a buggy arithmetic circuit, the vanishing monomials may propagate to the remainder because of a bug. To illustrate this phenomenon by an example, we show the backward rewriting process for a buggy 2-bit adder (cf. Fig. 8.1b):

$$SP \xrightarrow{P_{g_1}} SP_1 := 4w_1 + 4w_4 \boxed{-4w_1w_4} + 2z_1 + z_0 - (2a_1 + a_0 + 2b_1 + b_0),$$

$$SP_1 \xrightarrow{P_{g_2}} SP_2 := 4w_1 + 4w_2w_3 \boxed{-4w_1w_2w_3} + 2z_1 + z_0 - (2a_1 + a_0 + 2b_1 + b_0),$$

$$SP_2 \xrightarrow{P_{g_3}} SP_3 := 4w_1 \boxed{-4w_1w_2w_3} + 2w_2 + 2w_3 + z_0 - (2a_1 + a_0 + 2b_1 + b_0),$$

$$SP_3 \xrightarrow{P_{g4}} SP_4 := 2a_1 + 2b_1 - 4a_1b_1 \boxed{-4(a_1 + b_1 - a_1b_1)w_2w_3}$$
$$+ 2w_2 + 2w_3 + z_0 - (a_0 + b_0),$$

$$SP_4 \xrightarrow{P_{g5}} SP_5 := 4a_1 + 4b_1 - 8a_1b_1 \boxed{-4(a_1 + b_1 - 2a_1b_1)w_3}$$
$$+ 2w_3 + z_0 - (a_0 + b_0),$$

$$SP_5 \xrightarrow{P_{g6}} SP_6 := 4a_1 + 4b_1 - 8a_1b_1 \boxed{-4(a_1 + b_1 - 2a_1b_1)a_0b_0}$$
$$+ 2a_0b_0 + z_0 - (a_0 + b_0),$$

$$SP_6 \xrightarrow{P_{g7}} r := 4a_1 + 4b_1 - 8a_1b_1 \boxed{-4(a_1 + b_1 - 2a_1b_1)a_0b_0}. \quad (8.1)$$

The remainder r of backward rewriting is not equal to zero in Eq. (8.1). Thus, the circuit is buggy. The generated remainder consists of two parts. The first part $4a_1 + 4b_1 - 8a_1b_1 = 4 \times (a_1 + b_1 - 2a_1b_1)$ is composed of three monomials that originate from the difference in the buggy and correct gate polynomials at gate g_4 (see Fig. 8.1a and b):

$$P_{buggy} - P_{correct} = P_{OR} - P_{AND} = a_1 + b_1 - 2a_1b_1. \quad (8.2)$$

However, the second part of the remainder $-4(a_1 + b_1 - 2a_1b_1)a_0b_0$ (shown in the dashed box) is a part of the vanishing monomial propagated to the remainder due to the bug presence. If we apply the approach from [34] to the just discussed example, it fails. The reason is that for the vanishing monomials there will be no counterpart monomials when executing the forward rewriting and backward rewriting as presented in [34]. Furthermore, the SCA-based method from [33] also fails. This method extracts the difference polynomial per gate and compares it with the remainder. However, this is not possible if a vanishing monomial appears in the remainder. Finally, note that in all the experiments in both papers, only architecturally simple adders and multipliers are considered. The methods of [33, 34] cannot be employed for debugging architecturally complex multipliers where vanishing monomials are generated during backward rewriting. It is a serious shortcoming that limits their application to the verification of structurally simple multipliers.

8.3.2 Blow-up During the Verification of Buggy Circuits

If there is a bug close to the primary outputs of a large arithmetic circuit, a blow-up happens in the number of monomials during backward rewriting, and hence, SCA-based verification fails. The reason is that a buggy gate adds several monomials, i.e., the difference between buggy and correct gate polynomials, to the process of backward rewriting. These monomials create big peaks in the size of intermediate

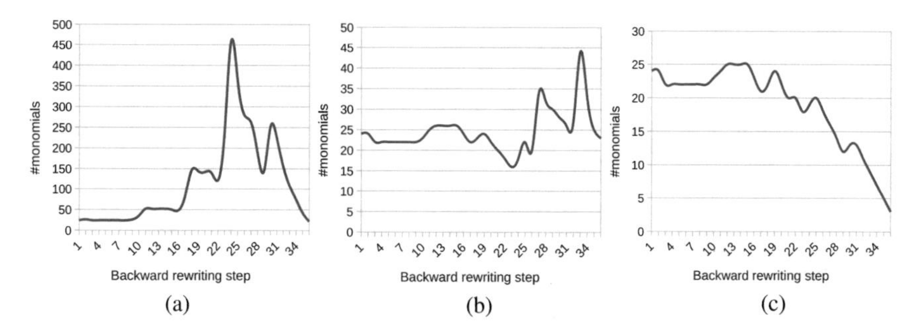

Fig. 8.2 Size of intermediate polynomials SP_i during the verification of a 4×4 buggy $SP \circ AR \circ RC$ multiplier. (**a**) Bug close to primary outputs. (**b**) Bug in middle. (**c**) Bug close to primary inputs

polynomials in the subsequent steps of the substitution when moving toward the inputs. Figure 8.2 shows the size of intermediate polynomials SP_i during the verification of a buggy 4-bit $SP \circ AR \circ RC$ multiplier when the bug is inserted close to the primary outputs (Fig. 8.2a), in the middle (Fig. 8.2b), and close to the primary inputs (Fig. 8.2c). If the bug is close to the primary outputs, the size of intermediate polynomials grows drastically, starting from the first steps. When the bug is in the middle of the circuit, we still observe considerable growth starting from the intermediate steps of backward rewriting. On the other hand, a bug close to the primary inputs does not cause a problem since the global backward rewriting finishes before the bug can generate a large number of extra monomials.

In a large multiplier with a bug close to the primary outputs, the size of the final remainder is very big and impossible to obtain. The SCA-based debugging methods in [33, 34, 74] require the remainder to detect and fix bugs. Therefore, they are not applicable when a bug occurs close to the primary outputs of a multiplier.

To overcome these limitations, we take advantage of both SAT and SCA in our approach. An overview is presented in the next section.

8.4 Proposed Debugging Method

In this section, we first give an overview of our complete debugging flow. Subsequently, we explain the three phases, i.e., verification, localization, and fixing, in detail.

8.4.1 Overview

Algorithm 6 shows the pseudocode of our proposed approach. Before we go into the details, it can be seen that our approach consists of three phases: *Verification*,

Algorithm 6 Proposed debugging method

Input: Arithmetic circuit C, Golden circuit C_G
Output: Correct circuit

```
 1: VerifyWithParallelSAT_SCA (C,C_G)                         ▷ Verification
 2: if C is bug-free then return C
 3: else
 4:     SG ← ExtractSGWithSAT (C, C_G)                        ▷ Localization
 5:     i ← 0; v ← ∅
 6:     while Size(SG) > 1 and i < 10 do
 7:         v ← GenerateTestvectorWithSAT (C, C_G, v)
 8:         C_X ← InsertXOR (C, SG, v)
 9:         SG ← RefineSGWithSCA(C_X)
10:         if SG has not changed then i ← i + 1
11:         else   i ← 0
12:     C_F ← FixWithParallelSAT_SCA (C, C_G, SG)             ▷ Fixing
13: return C_F
```

Localization, and *Fixing*. In each phase, we employ SAT and SCA for the individual subtasks as both have pros and cons. We explain the underlying decisions with regard to the chosen method. We provide a summary up-front in Table 8.1. The first column gives the name of the phase. The second column shows the subtask, if applicable. The third column distinguishes per phase/subtask between SCA and SAT. The fourth to seventh columns define whether the circuit is bug-free or not. In case of a bug, we subdivide the circuit into three regions: I, II, and III, which define the depth of the bug seen from the inputs (so III means a deep bug close to the primary outputs). Note that "+" means the respective method gives a result, and "−" that it fails. Finally, in the rightmost column, we show the conclusion that can be drawn. It has been finally implemented in our approach. In the following sections, we explain each phase in detail.

8.4.2 *Verification*

The first phase of our proposed approach is verification (see Lines 1–2 in Algorithm 6). In this phase, we want to determine whether an arithmetic circuit (particularly a structurally complex multiplier) is correct or not. SAT is very fast in disproving buggy circuits. If we create a miter using our buggy circuit and a reference (golden) circuit, a SAT-solver usually quickly finds the input values for which the CNF of the miter is satisfiable and returns them as a counter-example. However, SAT fails (times out) when the circuit is correct. In contrast, SCA is one of the best approaches for verifying bug-free arithmetic circuits. We showed in the previous chapter that our SCA-based verifier RevSCA-2.0 proves the correctness of a wide variety of large multipliers, including structurally complex designs. Nevertheless, the performance of SCA-based methods is poor when there is a bug close to the primary outputs (see previous discussion in Sect. 8.3). In order to take

Table 8.1 Applicability of SCA and SAT in different phases of debugging

Phase	Subtask	Method	Bug level				Conclusion
			Bug-free	I	II	III	
Verification		SCA	+	+	–	–	Using SAT and SCA in parallel (SAT for buggy and SCA for correct circuits)
		SAT	–	+	+	+	
Localization	Extracting initial suspicious gates	SCA		+	–	–	Using SAT
		SAT		+	+	+	
	Generating test-vectors	SCA		+	–	–	Using SAT
		SAT		+	+	+	
	Refining suspicious gates	SCA		++	++	++	Using SCA because it is faster
		SAT		+	+	+	
Fixing (correct fix/incorrect fix)		SCA		+/+*	+/–	+/–	Using SAT and SCA in parallel (SAT when fix does not work and SCA for proving correctness)
		SAT		–/+	–/+	–/+	

+ Applicable, ++ Applicable and fast, – Not applicable
*The sign before the slash (after the slash) describes the applicability of the method when the fix at the candidate location is correct (incorrect)

advantage of both SAT and SCA, we run them in parallel in our approach. When we obtain a result from one of the methods, we terminate the other one. As a result, buggy and bug-free circuits can be verified in an acceptable time.

Please note that since REVSCA-2.0 works on the AIG representation of a multiplier, we convert the gate-level description into AIG before the verification.

8.4.3 *Localization*

The second phase of our proposed method is the localization (see Lines 4–11 in Algorithm 6). The goal of this phase is to extract candidates for the location of the bug in the circuit. In the first step, an initial set of suspicious buggy gates SG is extracted (Line 4). Next, a test-vector is generated for the buggy circuit that activates the bug (Line 7). Subsequently, XOR gates are inserted just after each suspicious gate, and the test-vector is applied to the primary inputs (Line 8). Finally, the suspicious gates set is refined by backward rewriting and evaluating the remainder (Line 9). This process continues iteratively until either the size of the suspicious gates set SG reduces to 1, or the SG does not change after 10 iterations. In the following, we detail each step and explain for which subtask we employ which method, i.e., SCA or SAT (remember to see also Table 8.1 for a summary).

8.4.3.1 Extracting Initial Suspicious Gates

Arithmetic circuits, particularly multipliers, usually consist of many logic gates. Therefore, if there is a bug in the circuit, the size of the search space (i.e., the number of suspicious gates) will be large. Thus, a pre-process to reduce the size of search space is essential. To this end, we propose a pre-processing technique consisting of the three steps:

1. Identifying the primary outputs that are affected by the bug (i.e., there is an input vector such that the golden and buggy circuits differ).
2. Creating cones for these primary outputs based on the gates that are connected.
3. Determining the gates that are in the intersection of all cones; they form the initial set of suspicious gates.

This task can be mapped to a miter circuit for each output bit. In different studies, we observed that SAT performed very well, while an SCA-based solution only gives results for bugs in the circuit region I, i.e., bugs close to the primary inputs.

Consider the 2-bit adder example in Fig. 8.1b. The only affected output is Z_2. The cone for Z_2 is $C_2 = \{g_1, g_2, g_4, g_5, g_6\}$, which is also our initial suspicious gates set. Please note that if there are more than one affected output, the intersection of output cones creates the initial suspicious gate set.

8.4.3.2 Generating Test-Vectors

After the verification of an arithmetic circuit, a counter-example is available from the SAT-solver (SAT-solver result from Phase 1: Verification). This counter-example can be used as the initial test-vector since it presents the input values resulting in a "wrong" output value. However, we usually need more test-vectors to localize the bug in the design. To this end, we use blocking clauses and run the SAT-solver again to obtain a new test-vector if needed.

8.4.3.3 Insertion of XORs

The faulty gate in the circuit is among the gates in the suspicious gates set SG. Assume that t is a generated test-vector that exhibits the fault. Then, the bug has been activated in the circuit by t. In other words, the faulty gate has generated a "wrong" value at its output, which is the negation of the correct gate value—this assumption is valid since we consider the gate misplacement fault model. This "wrong" value is propagated through the gates in the output cone of the faulty gate and leads to a "wrong" value at the primary outputs of the circuit. Changing the output value of the faulty gate results in the correct value at the output of the circuit. Hence the problem can be formulated as follows: We look for the gates which negating their outputs rectifies the final result. As the XOR gate fulfills this property, we use it as follows: Assume that $g_1, \ldots, g_n$ are the suspicious gates, and t is a test-vector. We apply t to the primary inputs of the circuit and insert $x_1, \ldots, x_n$ that are XOR gates just after the suspicious gates. One input of each XOR gate is connected to the output of a suspicious gate, and the other XOR input becomes a new free input. We name all these inputs $s_1, \ldots, s_n$ and call them *selectors* in the following. The problem is now to find a *selector* by setting it to 1 (all others to 0), such that the final output of the circuit becomes correct.

Consider again the 2-bit buggy adder circuit in Fig. 8.1b. Recall that the initial set of suspicious gates is $\{g_1, g_2, g_4, g_5, g_6\}$. If $t_1 = 1000$ (i.e., $a_1 = 1, b_1 = 0, a_0 = 0, b_0 = 0$) is a test-vector, the new circuit after applying t_1 to the primary inputs and inserting XOR gates just after each suspicious gate is depicted in Fig. 8.1c.

8.4.3.4 Refining Suspicious Gates

The goal of this subtask is to refine the set of suspicious gates SG. In the following, we propose an SCA-based approach for this subtask, since it is faster than the SAT-based formulation (empirically shown in the experiments).

Based on the given test-vector t, we recompute the specification polynomial as follows: We have now concrete input values from t that are applied to the original specification polynomial. Please note that the inputs of the new circuit (current problem instance) are only the selectors. The gate polynomials of the 2-bit buggy adder with the previously inserted XORs and the test-vector $t_1 = 1000$ are

$$
\begin{aligned}
&P_{x_5} := Z_2 - s_5 - p_5 + 2s_5p_5, && P_{x_2} := w_2 - s_2 - p_2 + 2s_2p_2,\\
&P_{g_1} := p_5 - w_1 - w_4 + w_1w_4, && P_{x_3} := w_3 - s_3 - p_3 + 2s_3p_3,\\
&P_{x_4} := w_4 - s_4 - p_4 + 2s_4p_4, && P_{g_4} := p_1 - 1,\\
&P_{g_2} := p_4 - w_2w_3, && P_{g_5} := p_2 - 1,\\
&P_{g_3} := Z_1 - w_2 - w_3 + 2w_2w_3, && P_{g_6} := p_3 - 0,\\
&P_{x_1} := w_1 - s_1 - p_1 + 2s_1p_1, && P_{g_7} := Z_0 - 0.
\end{aligned} \tag{8.3}
$$

Now, backward rewriting is performed. The resulting remainder is different from 0 and only depends on the selectors. Before showing backward rewriting for the concrete example, two important points should be noticed: (1) the terms containing $s_m \times s_n$ (i.e., multiplication of selectors) are reduced to 0 during backward rewriting because only one of the selectors should be equal to 1; (2) due to the fact that we are dealing with a 2-bit adder, the specification polynomial and subsequently all the polynomials during backward rewriting should be modulo 2^{2+1} because the maximum output size for the addition of two n-bit numbers gives $n + 1$ bits. For an $n \times n$ multiplier, the calculations are always modulo 2^{2n} since the output always has $2n$ bits.

The backward rewriting steps for our running example are as follows:

$$
\begin{aligned}
&SP \xrightarrow{P_{x_5}} SP_1 := 4s_5 + 4p_5 + 2z_1 + z_0 - 2,\\
&SP_1 \xrightarrow{P_{g_1}} SP_2 := 4s_5 + 4w_1 + 4w_4 - 4w_1w_4 + 2z_1 + z_0 - 2,\\
&\dots\\
&SP_{10} \xrightarrow{P_{g_6}} SP_{11} := 4s_1 - 2s_2 - 2s_3 + 4s_5 + z_0 + 4,\\
&SP_{11} \xrightarrow{P_{g_7}} \boxed{r := 4s_1 - 2s_2 - 2s_3 + 4s_5 + 4}.
\end{aligned} \tag{8.4}
$$

To rectify the 2-bit adder circuit, the remainder $4s_1 - 2s_2 - 2s_3 + 4s_5 + 4$ should become 0. So, we should find all possible combinations for the selectors (one-hot encoding) such that $r = 0$. We get

$$
\begin{aligned}
&s_1 = 1, && s_2 = 0, && s_3 = 0, && s_4 = 0, && s_5 = 0 && \Longrightarrow && \boxed{\mathrm{r} = 8 \bmod 8 = 0},\\
&s_1 = 0, && s_2 = 1, && s_3 = 0, && s_4 = 0, && s_5 = 0 && \Longrightarrow && r = 2,\\
&s_1 = 0, && s_2 = 0, && s_3 = 1, && s_4 = 0, && s_5 = 0 && \Longrightarrow && r = 2,\\
&s_1 = 0, && s_2 = 0, && s_3 = 0, && s_4 = 1, && s_5 = 0 && \Longrightarrow && r = 4,\\
&s_1 = 0, && s_2 = 0, && s_3 = 0, && s_4 = 0, && s_5 = 1 && \Longrightarrow && \boxed{\mathrm{r} = 8 \bmod 8 = 0}.
\end{aligned} \tag{8.5}
$$

As can be seen in Eq. (8.5), when setting s_1 or s_5 to 1, the remainder becomes 0. Hence, the suspicious gates set is reduced to $\{g_4, g_1\}$ whose outputs are connected to x_1 and x_5, respectively. In the next iteration, XOR gates are inserted only after these two suspicious gates. Then, a new test-vector is generated and applied to the primary inputs, and suspicious gates are refined. Nevertheless, in this concrete example, the suspicious gates set cannot be further reduced even after applying all test-vectors. In large arithmetic circuits, the number of test-vectors is extremely large. Therefore, in order to avoid repeating subtasks 2, 3, and 4 for all existing test-vectors, we use the termination criteria of 10 iterations for the while-loop in Line 6 of Algorithm 6. In other words, if the size of the suspicious gates set does not change after 10 iterations, then the suspicious gates set is sent to the fixing phase.

8.4.4 Fixing

The final phase of our proposed method is Fixing (see Lines 12–13 in Algorithm 6). Based on the extracted candidates in the localization phase, we can create a list of potential gate replacements. For example, if we assume that the library for creating arithmetic circuits consists of the basic logic gates $\{AND, OR, XOR, NOT\}$, then there are two possible gate replacements for each candidate. To find the correct gate replacement, we first choose one of the changes from the list and apply it to the circuit. Then, we perform the parallel verification using SCA and SAT (see Sect. 8.4.2). Therefore, after gate replacement, if the circuit is still buggy, the SAT-based verification returns a counter-example, and we continue with the next possible replacement. Otherwise, if the circuit can be fixed with the current gate replacement, the SCA-based verification successfully proves it. Finally, the correct circuit after fixing is returned.

Considering again the running 2-bit buggy adder example of Fig. 8.1b from the localization phase, we know that g_4 and g_1 are the final suspicious gates. The corresponding list of gate replacements that may fix the bug is as follows:

$$\begin{aligned} &g_4(OR) \rightarrow g_4(XOR), \qquad &g_1(OR) \rightarrow g_1(XOR),\\ &g_4(OR) \rightarrow g_4(AND), \qquad &g_1(OR) \rightarrow g_1(AND). \end{aligned} \tag{8.6}$$

First, g_4 is converted into an XOR gate, and the circuit is verified. Because the circuit is still buggy, a counter-example is returned. When applying the second change (converting g_4 into an AND gate) and verifying the circuit, the final remainder of SCA-based verification becomes 0, and hence, we have found the fix.

8.5 Experimental Results

Our debugging approach has been implemented in C++. We take advantage of our SCA-based verifier REVSCA-2.0 during the verification, localization, and fixing phases. All experiments have been carried out on an Intel Xeon E3-1270 v3 with 3.50 GHz and 32 GByte of main memory. In order to evaluate the efficiency of our combined SCA and SAT approach, we consider different structurally complex multiplier architectures generated by AOKI [4]. The abbreviations for the used architectures are found in Table 2.2. We also used MiniSat v1.14 [29] for SAT-solving in our experiments.

The results of applying debugging methods to different types of multipliers are reported in Table 8.2. Please note that the *Time-Out* (T.O.) has been set to 24 h. In addition, *Not Applicable* (N.A.) means that the method cannot be employed for debugging the multiplier due to its limitations. The first column of Table 8.2 **Benchmark** shows the type of the multiplier. The second column **Size** gives the number of input bits. The third column **Bug** lists whether the circuit is bug-free, or the stage where the bug has been inserted randomly.

The results of the verification phase are reported in the fourth column **Verification**, which consists of the five following subcolumns: While *SCA* and *SAT* refer to our SCA-based verifier REVSCA-2.0 and MiniSat, respectively, *Comm.* reports the results of a commercial formal verification tool. Next, the run-time of our integrated approach is given in subcolumn *Ours*, and finally, *Imp.* presents the improvement of our approach compared to the commercial tool. As can be seen, pure SCA-based verification only works when there is no bug in the circuit, or the bug is in the first stage (i.e., PPG) of the design. In contrast, pure SAT-based verification times out for bug-free circuits. The commercial tool also times out for bug-free multipliers bigger than 16×16. On the other hand, our integrated verification method (Sect. 8.4.2) can verify bug-free multipliers as well as buggy circuits when the bug is in any stage of the design. Our verification method is up to 550 times faster than the commercial tool.

The fifth column **Localization** shows the run-times of the localization phase. We have compared our method against the SAT-based debugging [79]. While the SAT-based debugging is able to compute the set of fault candidates for all three stages, our method is faster for all benchmarks. The respective improvements are listed in the third subcolumn. As can be seen, we achieve improvements of up to a factor of 145.

The experimental results for the fixing phase are reported in the sixth column **Fixing**. As can be seen, the pure SCA-based fixing method is only able to fix bugs in the first stage (PPG) of the multiplier. The SAT-based fixing fails for all the cases because when a correct gate replacement is considered, this method cannot verify the bug-free circuit. The experimental results confirm that our fixing method can fix the bugs at any stage of the design.

Table 8.2 Results of debugging different types of multipliers (run-times in seconds)

Benchmark	Size	Bug	Verification					Localization			Fixing			Overall	SOTA	
			SCA	SAT	Comm.	Ours	Imp.	SAT [79]	Ours	Imp.	SCA	SAT	Ours	Ours	[34]	[33]
SP-CT-BK	16 × 16	Bug-free	0.09	T.O.	55.00	0.10	550.00x							0.1		
		Stage 1	0.09	0.04	0.07	0.05	1.40x	24.6	0.6	41.0x	0.28	T.O.	0.15	0.8	N.A.	N.A.
		Stage 2	T.O.	0.02	0.07	0.02	3.50x	43.8	0.6	73.0x	T.O.	T.O.	0.16	0.8	N.A.	N.A.
		Stage 3	T.O.	0.03	0.06	0.03	2.00x	11.0	0.7	15.7x	T.O.	T.O.	0.15	0.9	N.A.	N.A.
BP-WT-CL	16 × 16	Bug-free	0.15	T.O.	61.00	0.16	381.25x							0.16		
		Stage 1	0.17	0.02	0.06	0.02	3.00x	32.0	0.8	40.0x	0.49	T.O.	0.22	1.0	N.A.	N.A.
		Stage 2	T.O.	0.03	0.06	0.04	1.50x	55.8	0.8	68.8x	T.O.	T.O.	0.24	1.1	N.A.	N.A.
		Stage 3	T.O.	0.15	0.17	0.17	1.00x	7.2	1.0	7.2x	T.O.	T.O.	0.45	1.6	N.A.	N.A.
BP-CT-BK	16 × 16	Bug-free	0.13	T.O.	63.00	0.15	420.00x							0.15		
		Stage 1	0.13	0.03	0.06	0.03	2.00x	13.1	0.7	18.7x	0.42	T.O.	0.28	0.9	N.A.	N.A.
		Stage 2	T.O.	0.02	0.05	0.02	2.50x	14.9	0.7	25.6x	T.O.	T.O.	0.24	1.0	N.A.	N.A.
		Stage 3	T.O.	0.03	0.07	0.03	2.33x	10.7	0.6	17.8x	T.O.	T.O.	0.25	0.9	N.A.	N.A.
SP-WT-CL	32 × 32	Bug-free	0.96	T.O.	T.O.	1.17	–							1.17		
		Stage 1	0.98	0.20	0.38	0.22	1.73x	115.6	7.5	15.4x	2.38	T.O.	1.33	9.1	N.A.	N.A.
		Stage 2	T.O.	0.16	0.41	0.20	2.05x	1208.0	8.3	145.5x	T.O.	T.O.	1.28	9.8	N.A.	N.A.
		Stage 3	T.O.	0.12	0.35	0.14	2.50x	165.4	7.0	23.6x	T.O.	T.O.	1.25	8.4	N.A.	N.A.
BP-AR-RC	32 × 32	Bug-free	0.89	T.O.	T.O.	0.95	–							0.95		
		Stage 1	0.96	0.14	0.24	0.15	1.60x	208.7	6.7	31.1x	3.87	T.O.	1.05	7.9	N.A.	N.A.
		Stage 2	T.O.	0.10	0.31	0.13	2.38x	89.3	6.1	14.6x	T.O.	T.O.	1.10	7.3	N.A.	N.A.
		Stage 3	T.O.	0.08	0.30	0.09	3.33x	60.8	5.9	10.3x	T.O.	T.O.	1.08	7.1	N.A.	N.A.

(continued)

Table 8.2 (continued)

Benchmark	Size	Bug	Verification					Localization			Fixing			Overall	SOTA	
			SCA	SAT	Comm.	Ours	Imp.	SAT [79]	Ours	Imp.	SCA	SAT	Ours	Ours	[34]	[33]
BP-WT-CL	32 × 32	Bug-free	1.38	T.O.	T.O.	1.45	–							1.45		
		Stage 1	1.41	0.08	0.23	0.08	2.87x	291.2	6.7	43.5x	4.21	T.O.	1.62	8.4	N.A.	N.A.
		Stage 2	T.O.	0.11	0.27	0.12	2.25x	228.2	10.5	21.1x	T.O.	T.O.	1.54	12.2	N.A.	N.A.
		Stage 3	T.O.	0.45	0.32	0.56	0.57x	148.1	4.3	34.4x	T.O.	T.O.	1.65	6.51	N.A.	N.A.
SP-WT-CL	64 × 64	Bug-free	15.17	T.O.	T.O.	15.93	–							15.93		
		Stage 1	15.96	1.14	12.66	1.22	10.38x	1107.4	70.9	3.23x	46.80	T.O.	17.68	89.8	N.A.	N.A.
		Stage 2	T.O.	3.86	77.00	4.14	18.60x	1745.6	65.1	6.97x	T.O.	T.O.	19.20	88.4	N.A.	N.A.
		Stage 3	T.O.	0.57	4.46	0.63	7.08x	1047.9	79.5	2.02x	T.O.	T.O.	17.57	97.7	N.A.	N.A.
SP-CT-BK	64 × 64	Bug-free	22.68	T.O.	T.O.	22.88	–							22.88		
		Stage 1	22.90	1.79	42.47	1.91	22.24x	1192.3	83.3	14.3x	132.01	T.O.	72.12	157.3	N.A.	N.A.
		Stage 2	T.O.	0.42	22.05	0.46	47.93x	677.7	71.6	9.5x	T.O.	T.O.	69.21	141.3	N.A.	N.A.
		Stage 3	T.O.	0.88	9.56	0.94	10.17x	3612.0	77.8	46.4x	T.O.	T.O.	73.54	152.3	N.A.	N.A.
BP-WT-CL	64 × 64	Bug-free	20.91	T.O.	T.O.	21.38	–							21.38		
		Stage 1	21.04	0.94	2.05	0.99	2.07x	1323.1	72.2	18.3x	64.12	T.O.	23.87	97.1	N.A.	N.A.
		Stage 2	T.O.	0.43	2.61	0.53	4.92x	1669.0	59.2	28.2x	T.O.	T.O.	22.44	82.2	N.A.	N.A.
		Stage 3	T.O.	2.28	14.85	2.40	6.19x	698.6	64.9	10.8x	T.O.	T.O.	26.14	93.4	N.A.	N.A.

The overall run-time of our proposed method is reported in the seventh column **Overall**. It gives the sum of run-times for each phase, i.e., verification, localization, and fixing.

The eighth column shows the debugging results of the *state-of-the-art* (SOTA) SCA-based methods [34] and [33]. Both methods cannot be used to debug the considered structurally complex multipliers due to either appearance of vanishing monomials in the final remainder or the explosion during the verification (see discussions and examples in Sect. 8.3).

8.6 Conclusion

In this chapter, we proposed a complete debugging flow to verify a multiplier, localize bugs, and fix them. We first introduced our fault model known as gate misplacement. This fault model is very popular since it covers many bugs in real-world designs. A wrong gate can occur in a circuit in different design phases such as the generation, synthesis, or optimization.

Second, we highlighted the limitations of SCA-based methods for debugging. The existing techniques only work for structurally simple multipliers and under certain conditions. The two critical obstacles are the appearance of vanishing monomials in the remainder and the explosion in the number of monomials when the bug is close to the primary outputs.

Then, we introduced our proposed debugging approach, consisting of three phases: verification, localization, and fixing. We explained the advantages and disadvantages of using SCA and SAT for each phase. Thus, we chose the combinations of SAT and SCA in order to carry out each phase as fast as possible.

Finally, we evaluated the efficiency of our method in debugging a wide variety of multiplier architectures with different sizes. Our method can successfully verify a multiplier, localize the bug, and fix it. The experimental results confirm that our debugging method is the fastest approach when it comes to debugging structurally complex multipliers.

Chapter 9
Conclusion and Outlook

This chapter concludes this book by referring to the initial research objectives in Chap. 1 and outlines further research based on the proposed contributions.

9.1 Conclusion

Back in 1970, an Intel 4004 processor had 2,250 transistors. It could only support a limited number of instructions, and it was working at a very low frequency. However, the digital circuits nowadays are much larger, sometimes even consisting of billions of transistors. Moreover, they are usually designed based on sophisticated algorithms, leading to fast but complex architectures. The big size and the high complexity of modern digital circuits make them extremely error-prone during different design phases. The implementation and production of buggy circuits can cause a catastrophe, resulting in huge financial losses and endangering lives. Consequently, formal verification is an important task to ensure the correctness of a digital circuit.

Formal verification of arithmetic circuits is one of the most challenging problems in the verification community. Although SAT-based and BDD-based verification reported good results for integer adders, the formal verification of integer multipliers and dividers remained unsolved for a long time. In the last six years, the SCA-based verification methods achieved many successes in proving the correctness of structurally simple multipliers. The proposed techniques can verify a very large multiplier in a few seconds. However, they either totally fail or support a limited set of benchmarks when it comes to verifying structurally complex multipliers.

This book addressed the challenging task of verifying structurally complex multipliers. The focus of this book was particularly on achieving two objectives: (1) extending the basic SCA-based verification method to overcome the challenges

A. Mahzoon et al., *Formal Verification of Structurally Complex Multipliers*,
https://doi.org/10.1007/978-3-031-24571-8_9

of proving the correctness of structurally complex multipliers and (2) combining SCA and SAT to localize and fix bugs in structurally complex multipliers.

Chapters 3, 4, 5, 6, and 7 were dedicated to accomplishing the first objective. In Chap. 3, the definitions of structurally simple and complex multipliers were given. Then, the challenges of verifying structurally complex multipliers were illustrated by different experiments. The huge number of monomials during the backward rewriting of these multipliers leads to an explosion and, subsequently, the verification failure. Three techniques, i.e., local vanishing monomials removal, reverse engineering, and dynamic backward rewriting, were proposed in Chaps. 4, 5, and 6 to extend SCA-based verification and avoid the explosion. Finally, Chap. 7 introduced our SCA-based verifier REVSCA-2.0. The three proposed techniques are integrated into REVSCA-2.0 to verify structurally complex multipliers. The experimental results confirmed that REVSCA-2.0 can prove the correctness of a wide variety of structurally complex multipliers, including both clean and dirty optimized architectures. Currently, REVSCA-2.0 is the most robust verification tool when it comes to verifying optimized multipliers. It supports the verification of several optimized and technology-mapped architectures, including industrial designs, while the other state-of-the-art verification methods totally fail.

Chapter 8 was dedicated to achieving the second objective. After highlighting the limitations of the pure SCA-based methods, a complete debugging flow based on SCA and SAT was proposed in this chapter. The flow consists of three phases, i.e., verifying the circuit, localizing the bug, and fixing it. Then, the advantages of using SCA and SAT for each phase were explained. The experimental results showed that our proposed approach can successfully debug a wide variety of structurally complex multipliers, while the other SCA-based debugging methods fail.

9.2 Outlook

The proposed contributions presented in this book can be used as the foundation for further work.

Chapter 4 proposed a vanishing monomials removal technique to avoid explosions caused by vanishing monomials during backward rewriting. This technique can remove vanishing monomials originating from half-adders in a multiplier. However, the concept of vanishing monomials can be further generalized: assume for a pair of signals a and b, both a and b cannot be 1 at the same time. If a and b converge to a node, the vanishing monomials might appear during backward rewriting. This generalization would be useful to detect and remove vanishing monomials in highly optimized multipliers. However, detecting CNCs in highly optimized multipliers is a challenging task, which has to be addressed.

Chapter 5 proposed a reverse engineering technique to identify half-adders, full-adders, and (4:2) compressors. The proposed technique can be extended to larger atomic blocks, e.g., (5:3) compressors and (7:3) counters. The detection of these blocks makes REVSCA-2.0 applicable to a wider variety of multipliers. However, it requires enumerating larger cuts, which is a challenging and time-consuming task. Moreover, reconstructing atomic blocks in highly optimized multipliers can be beneficial for the verification of these multipliers.

Chapter 6 proposed a dynamic backward rewriting technique to control the size of intermediate polynomials SP_i during backward rewriting. Our proposed method restores SP_i to its previous state if a sudden increase happens in its size. It is essential to increase the depth of backtracking in order to restore SP_i more than one step. This is potentially useful for the verification of highly optimized multipliers. However, it requires storing the copies of several intermediate polynomials, which is expensive in terms of memory usage and run-time. Overcoming these challenges is a step forward toward formal verification of highly optimized multipliers.

Chapter 7 introduced REVSCA-2.0 to verify structurally complex multipliers. Despite the success of REVSCA-2.0 in ensuring the correctness of a wide variety of clean and dirty multipliers, its space and time complexities are not investigated. Researchers have recently put a lot of efforts in calculating the upper-bound space and time complexities of verifying adders [22, 47, 48] and multipliers [5, 26]. These outcomes can help us to calculate the complexity of REVSCA-2.0 for various architectures and to see whether polynomial formal verification [25] can be achieved.

Chapter 8 proposed a complete debugging flow to localize and fix bugs in multipliers. The proposed approach uses gate misplacement as its fault model and supports the debugging of a single bug in the circuit. In order to support other fault models and multiple bugs, an extension of the debugging approach is required. It is a subject for future research.

In addition to integer multipliers, the proposed methods in this book can be used in verifying and debugging modular multipliers (see [55]), truncated multipliers, and dividers.

Appendix A
SCA-Verification Website

The SCA-based verification approaches proposed in this book are available on http://www.sca-verification.org. The website includes the explanation of each verification method, as well as the links to the GitHub pages, where users can find the tools and benchmarks. The latest version of REVSCA-2.0 introduced in Chap. 7 is also accessible via the website.

In addition to the verification approaches, the latest version of our multiplier generator GENMUL is now available on http://www.sca-verification.org/genmul. The webpage includes the link to the source code of GENMUL on GitHub, as well as the web-based version of the tool. In Fig. A.1, a screenshot of the web-based version of GENMUL is shown.

A. Mahzoon et al., *Formal Verification of Structurally Complex Multipliers*,
https://doi.org/10.1007/978-3-031-24571-8

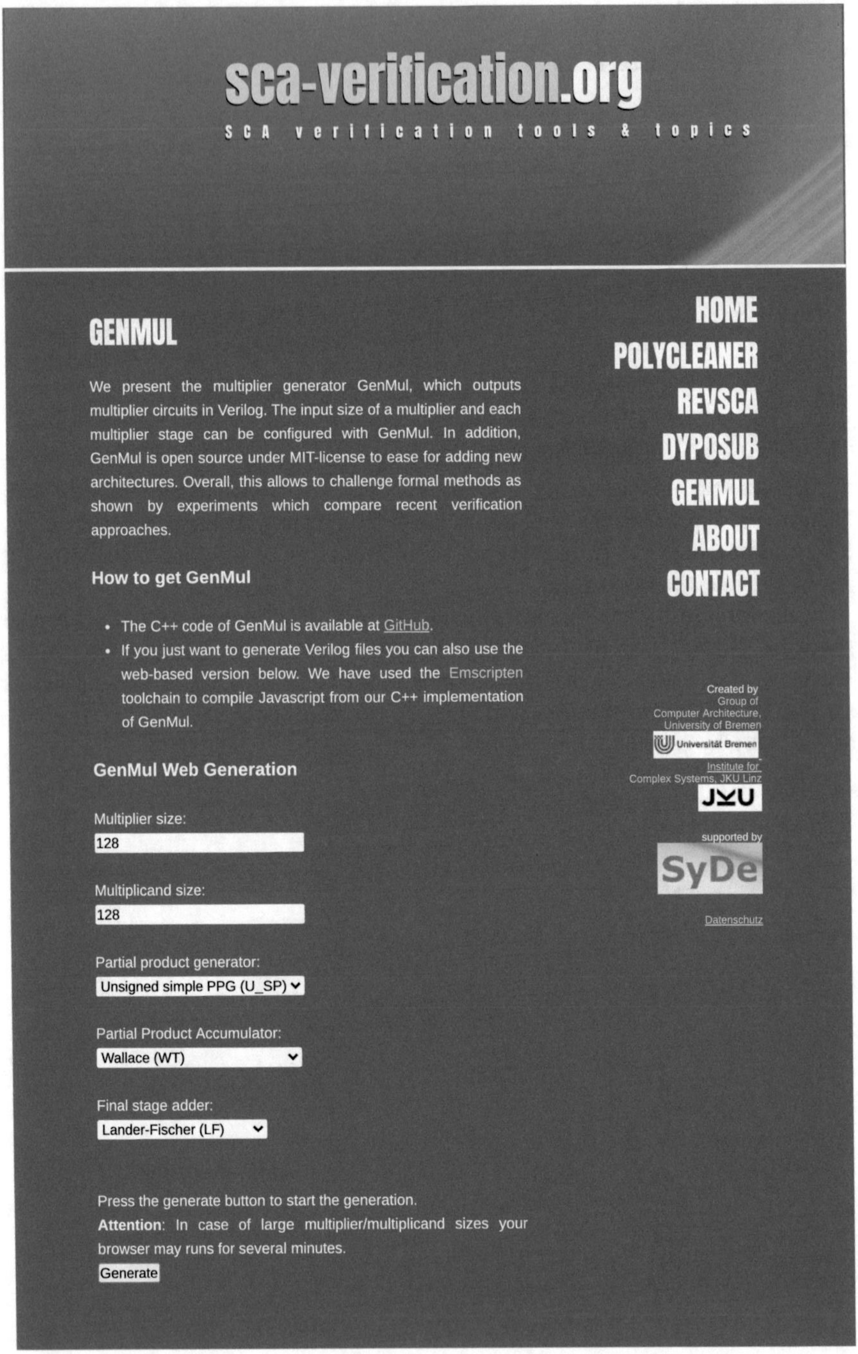

sca-verification.org

SCA verification tools & topics

HOME
POLYCLEANER
REVSCA
DYPOSUB
GENMUL
ABOUT
CONTACT

GENMUL

We present the multiplier generator GenMul, which outputs multiplier circuits in Verilog. The input size of a multiplier and each multiplier stage can be configured with GenMul. In addition, GenMul is open source under MIT-license to ease for adding new architectures. Overall, this allows to challenge formal methods as shown by experiments which compare recent verification approaches.

How to get GenMul

- The C++ code of GenMul is available at GitHub.
- If you just want to generate Verilog files you can also use the web-based version below. We have used the Emscripten toolchain to compile Javascript from our C++ implementation of GenMul.

GenMul Web Generation

Multiplier size:
128

Multiplicand size:
128

Partial product generator:
Unsigned simple PPG (U_SP)

Partial Product Accumulator:
Wallace (WT)

Final stage adder:
Lander-Fischer (LF)

Press the generate button to start the generation.
Attention: In case of large multiplier/multiplicand sizes your browser may runs for several minutes.
Generate

Created by
Group of
Computer Architecture,
University of Bremen
Universität Bremen
Institute for
Complex Systems, JKU Linz
JKU
supported by
SyDe
Datenschutz

Fig. A.1 GENMUL website

References

1. ABC: A system for sequential synthesis and verification (2018). https://people.eecs.berkeley.edu/~alanmi/abc/
2. W.W. Adamsm P. Loustaunau, *An Introduction to Gröbner Bases.* (American Mathematical Society, 1994)
3. L. Amaru, P.-E. Gaillardon, G. De Micheli, The EPFL combinational benchmark suite, in *International Workshop on Logic and Synthesis* (2015)
4. Arithmetic module generator based on ACG (2019). https://www.ecsis.riec.tohoku.ac.jp/topics/amg/i-amg
5. M. Barhoush, A. Mahzoon, R. Drechsler, Polynomial word-level verification of arithmetic circuits, in *ACM & IEEE International Conference on Formal Methods and Models for Codesign* (2021), pp. 1–9
6. M. Blum, H. Wasserman, Reflections on the Pentium division bug. IEEE Trans. Comput. **45**(4), 385–393 (1996)
7. A.D. Booth, A signed binary multiplication technique. Q. J. Mech. Appl. Math. **4**(2), 236–240 (1951)
8. R.E. Bryant, Graph-based algorithms for Boolean function manipulation. IEEE Trans. Comput. **35**(8), 677–691 (1986)
9. R.E. Bryant, On the complexity of VLSI implementations and graph representations of Boolean functions with application to integer multiplication. IEEE Trans. Comput. **40**(2), 205–213 (1991)
10. R.E. Bryant, Binary decision diagrams and beyond: enabling technologies for formal verification, in *International Conference on Computer-Aided Design* (1995), pp. 236–243
11. R.E. Bryant, Y.A. Chen, Verification of arithmetic circuits with binary moment diagrams, in *Design Automation Conference* (1995), pp. 535–541
12. R. Camposano, W. Rosenstiel, Synthesizing circuits from behavioural descriptions. IEEE Trans. Comput. Aided Design Circuits Syst. **8**(2), 171–180 (1989)
13. S. Chatterjee, A. Mishchenko, R.K. Brayton, Factor cuts, in *International Conference on Computer-Aided Design* (2006), pp. 143–150
14. M. Ciesielski, T. Su, A. Yasin, C. Yu, Understanding algebraic rewriting for arithmetic circuit verification: a bit-flow model. IEEE Trans. Comput. Aided Design Circuits Syst. **39**(6), 1346–1357 (2019)
15. M. Ciesielski, C. Yu, D. Liu, W. Brown, Verification of gate-level arithmetic circuits by function extraction, in *Design Automation Conference* (2015), pp. 52:1–52:6
16. J. Cong, C. Wu, Y. Ding, Cut ranking and pruning: enabling a general and efficient FPGA mapping solution, in *International Symposium on Field Programmable Gate Arrays* (1999), pp. 29–35

A. Mahzoon et al., *Formal Verification of Structurally Complex Multipliers*,
https://doi.org/10.1007/978-3-031-24571-8

17. D.A. Cox, J. Little, D. O'Shea, *Ideals Varieties and Algorithms* (Springer, Berlin, 1997)
18. B. Crothers, Intel's Sandy Bridge chipset flaw: the fallout
19. S. Disch, C. Scholl, Combinational equivalence checking using incremental SAT solving, output ordering, and resets, in *Asia and South Pacific Design Automation Conference* (2007), pp. 938–943
20. R. Drechsler, *Advanced Formal Verification* (Kluwer Academic Publishers, 2004)
21. R. Drechsler, *Formal System Verification: State-of the-Art and Future Trends* (Springer, Berlin, 2017)
22. R. Drechsler, PolyAdd: polynomial formal verification of adder circuits, in *IEEE Symposium on Design and Diagnostics of Electronic Circuits and Systems* (2021), pp. 99–104
23. R. Drechsler, B. Becker, S. Ruppertz, The K*BMD: a verification data structure. IEEE Design Test Comput. **14**(2), 51–59 (1997)
24. R. Drechsler, S. Höreth, Manipulation of *BMDs, in *Asia and South Pacific Design Automation Conference* (1998), pp. 433–438
25. R. Drechsler, A. Mahzoon, Polynomial formal verification: ensuring correctness under resource constraints, in *International Conference on Computer-Aided Design* (2022)
26. R. Drechsler, A. Mahzoon, Towards polynomial formal verification of complex arithmetic circuits, in *IEEE Symposium on Design and Diagnostics of Electronic Circuits and Systems* (2022)
27. R. Drechsler, D. Sieling, Binary decision diagrams in theory and practice. Int. J. Softw. Tools Technol. Transf. **3**(2), 112–136 (2001)
28. A. Edelman, The mathematics of the Pentium division bug. SIAM Rev. **39**(1), 54–67 (1997)
29. N. Eén, N. Sörensson, MiniSat (2008). http://minisat.se/
30. Emscripten (2019). https://emscripten.org
31. F. Farahmandi, B. Alizadeh, Gröbner basis based formal verification of large arithmetic circuits using Gaussian elimination and cone-based polynomial extraction. Microprocess. Microsyst. **39**(2), 83–96 (2015)
32. F. Farahmandi, B. Alizadeh, Z. Navabi, Effective combination of algebraic techniques and decision diagrams to formally verify large arithmetic circuits, in *IEEE Annual Symposium on VLSI* (2014), pp. 338–343
33. F. Farahmandi, P. Mishra, Automated test generation for debugging arithmetic circuits, in *Design, Automation and Test in Europe* (2016), pp. 1351–1356
34. S. Ghandali, C. Yu, D. Liu, W. Brown, M. Ciesielski, Logic debugging of arithmetic circuits, in *IEEE Annual Symposium on VLSI* (2015), pp. 113–118
35. E.I. Goldberg, M.R. Prasad, R.K. Brayton, Using SAT for combinational equivalence checking, in *Design, Automation and Test in Europe* (2001), pp. 114–121
36. U. Gupta, P. Kalla, I. Ilioaea, F. Enescu, Exploring algebraic interpolants for rectification of finite field arithmetic circuits with Gröbner bases, in *European Test Symposium* (2019), pp. 1–6
37. U. Gupta, P. Kalla, V. Rao, Boolean Gröbner basis reductions on finite field datapath circuits using the unate cube set algebra. IEEE Trans. Comput. Aided Design Circuits Syst. **38**(3), 576–588 (2019)
38. K. Hamaguchi, A. Morita, S. Yajima, Efficient construction of binary moment diagrams for verifying arithmetic circuits, in *International Conference on Computer-Aided Design* (1995), pp. 78–82
39. S. Höreth, R. Drechsler, Formal verification of word-level specifications, in *Design, Automation and Test in Europe* (1999), pp. 52–58
40. Z. Huang, L. Wang, Y. Nasikovskiy, A. Mishchenko, Fast Boolean matching based on NPN classification, in *International Conference on Field-Programmable Technology* (2013), pp. 310–313
41. D. Kaufmann, A. Biere, M. Kauers, Verifying large multipliers by combining SAT and computer algebra, in *Formal Methods in Computer-Aided Design* (2019), pp. 28–36
42. D. Kaufmann, A. Biere, M. Kauers, Incremental column-wise verification of arithmetic circuits using computer algebra. Formal Methods Syst. Design Int. J. **56**(1), 22–54 (2020)

43. A. Konrad, C. Scholl, A. Mahzoon, D. Große, R. Drechsler, Divider verification using symbolic computer algebra and delayed don't care optimization, in *Formal Methods in Computer-Aided Design* (2022)
44. I. Koren, *Computer Arithmetic Algorithms*, 2nd edn. (A. K. Peters, 2001)
45. A. Kuehlmann, F. Krohm, Equivalence checking using cuts and heaps, in *Design Automation Conference* (1997), pp. 263–268
46. A. Kuehlmann, V. Paruthi, F. Krohm, M.K. Ganai, Robust Boolean reasoning for equivalence checking and functional property verification. IEEE Trans. Comput. Aided Design Circuits Syst. **21**(12), 1377–1394 (2002)
47. A. Mahzoon, R. Drechsler, Late breaking results: polynomial formal verification of fast adders, in *Design Automation Conference* (2021), pp. 1376–1377
48. A. Mahzoon, R. Drechsler, Polynomial formal verification of prefix adders, in *Asian Test Symp.* (2021), pp. 85–90
49. A. Mahzoon, D. Große, R. Drechsler, Combining symbolic computer algebra and Boolean satisfiability for automatic debugging and fixing of complex multipliers, in *IEEE Annual Symposium on VLSI* (2018), pp. 351–356
50. A. Mahzoon, D. Große, R. Drechsler, PolyCleaner: clean your polynomials before backward rewriting to verify million-gate multipliers, in *International Conference on Computer-Aided Design* (2018), pp. 129:1–129:8
51. A. Mahzoon, D. Große, R. Drechsler, RevSCA: using reverse engineering to bring light into backward rewriting for big and dirty multipliers, in *Design Automation Conference* (2019), pp. 185:1–185:6
52. A. Mahzoon, D. Große, R. Drechsler, GenMul: generating architecturally complex multipliers to challenge formal verification tools, in *Recent Findings in Boolean Techniques*, ed. by R. Drechsler, D. Große (Springer, Berlin, 2021), pp. 177–191
53. A. Mahzoon, D. Große, R. Drechsler, RevSCA-2.0: SCA-based formal verification of non-trivial multipliers using reverse engineering and local vanishing removal. IEEE Trans. Comput. Aided Design Circuits Syst. **41**(5), 1573–1586 (2021)
54. A. Mahzoon, D. Große, C. Scholl, R. Drechsler, Towards formal verification of optimized and industrial multipliers, in *Design, Automation and Test in Europe* (2020), pp. 544–549
55. A. Mahzoon, D. Große, C. Scholl, A. Konrad, R. Drechsler, Formal verification of modular multipliers using symbolic computer algebra and Boolean satisfiability, in *Design Automation Conference* (2022)
56. A. Mishchenko, R.K. Brayton, S. Chatterjee, Boolean factoring and decomposition of logic networks, in *International Conference on Computer-Aided Design* (2008), pp. 38–44
57. A. Mishchenko, R.K. Brayton, S. Jang, V.N. Kravets, Delay optimization using SOP balancing, in *International Conference on Computer-Aided Design* (2011), pp. 375–382
58. A. Mishchenko, R.K. Brayton, J.R. Jiang, S. Jang, Scalable don't-care-based logic optimization and resynthesis, in *International Symposium on Field Programmable Gate Arrays* (2009), pp. 151–160
59. A. Mishchenko, S. Chatterjee, R.K. Brayton, DAG-aware AIG rewriting a fresh look at combinational logic synthesis, in *Design Automation Conference* (2006), pp. 532–535
60. A. Mishchenko, S. Cho, S. Chatterjee, R.K. Brayton, Combinational and sequential mapping with priority cuts, in *International Conference on Computer-Aided Design* (2007), pp. 354–361
61. Z. Navabi, *Verilog Digital System Design: Register Transfer Level Synthesis, Testbench, and Verification* (McGraw-Hill Professional, 2005)
62. P. Pan, C.-C. Lin, A new retiming-based technology mapping algorithm for LUT-based FPGAs, in *FPGAs for Custom Computing Machines* (1998), pp. 35–42
63. B. Parhami, *Computer Arithmetic—Algorithms and Hardware Designs* (Oxford University Press, Oxford, 2000)
64. P. Pourbeik, P.S. Kundur, C.W. Taylor, The anatomy of a power grid blackout-root causes and dynamics of recent major blackouts. IEEE Power Energy Mag. **4**(5), 22–29 (2006)
65. V. Pratt, Anatomy of the Pentium bug, in *TAPSOFT '95: Theory and Practice of Software Development* (1995. Springer, Berlin, Heidelberg), pp. 97–107

66. T. Pruss, P. Kalla, F. Enescu, Efficient symbolic computation for word-level abstraction from combinational circuits for verification over finite fields. IEEE Trans. Comput. Aided Design Circuits Syst. **35**(7), 1206–1218 (2016)
67. V. Rao, I. Ilioaea, H. Ondricek, P. Kalla, F. Enescu, Word-level multi-fix rectifiability of finite field arithmetic circuits, in *International Symposium on Quality Electronic Design* (2021), pp. 41–47
68. V. Rao, H. Ondricek, P. Kalla, F. Enescu, Algebraic techniques for rectification of finite field circuits, in *VLSI of System-on-Chip* (2021), pp. 1–6
69. V. Rao, H. Ondricek, P. Kalla, F. Enescu, Rectification of integer arithmetic circuits using computer algebra techniques, in *International Conference on Computer Design* (2021), pp. 186–195
70. S. Ray, A. Mishchenko, N. Eén, R.K. Brayton, S. Jang, C. Chen, Mapping into LUT structures, in *Design, Automation and Test in Europe* (2012), pp. 1579–1584
71. D. Ritirc, A. Biere, M. Kauers, Column-wise verification of multipliers using computer algebra, in *Formal Methods in Computer-Aided Design* (2017), pp. 23–30
72. D. Ritirc, A. Biere, M. Kauers, Improving and extending the algebraic approach for verifying gate-level multipliers, in *Design, Automation and Test in Europe* (2018), pp. 1556–1561
73. A. Rushton, *VHDL for Logic Synthesis* (John Wiley & Sons, 1998)
74. N.A. Sabbagh, B. Alizadeh, Arithmetic circuit correction by adding optimized correctors based on Grobner basis computation, in *European Test Symposium* (2021), pp. 1–6
75. A. Sayed-Ahmed, D. Große, U. Kühne, M. Soeken, R. Drechsler, Formal verification of integer multipliers by combining Gröbner basis with logic reduction, in *Design, Automation and Test in Europe* (2016), pp. 1048–1053
76. A. Sayed-Ahmed, D. Große, M. Soeken, R. Drechsler, Equivalence checking using Gröbner bases, in *Formal Methods in Computer-Aided Design* (2016), pp. 169–176
77. C. Scholl, A. Konrad, Symbolic computer algebra and SAT based information forwarding for fully automatic divider verification, in *Design Automation Conference* (2020), pp. 1–6
78. C. Scholl, A. Konrad, A. Mahzoon, D. Große, R. Drechsler, Verifying dividers using symbolic computer algebra and don't care optimization, in *Design, Automation and Test in Europe* (2021), pp. 1110–1115
79. A. Smith, A.G. Veneris, A. Viglas, Design diagnosis using Boolean satisfiability, in *Asia and South Pacific Design Automation Conference* (2004), pp. 218–223
80. M. Soeken, H. Riener, W. Haaswijk, G.D. Micheli, The EPFL logic synthesis libraries (2018). arXiv:1805.05121
81. M. Temel, W.A. Hunt, Sound and automated verification of real-world RTL multipliers, in *Formal Methods in Computer-Aided Design* (2021), pp. 53–62
82. M. Temel, A. Slobodová, W.A. Hunt, Automated and scalable verification of integer multipliers, in *Computer Aided Verification* (2020), pp. 485–507
83. E. Testa, M. Soeken, L. Amarù, G.D. Micheli, Reducing the multiplicative complexity in logic networks for cryptography and security applications, in *Design Automation Conference* (2019), pp. 1–6
84. G.S. Tseitin, *On the Complexity of Derivation in Propositional Calculus* (Springer, Berlin, Heidelberg, 1983), pp. 466–483
85. S. Vasudevan, V. Viswanath, R.W. Sumners, J.A. Abraham, Automatic verification of arithmetic circuits in RTL using stepwise refinement of term rewriting systems. IEEE Trans. Comput. **56**(10), 1401–1414 (2007)
86. A. Veneris, I.N. Hajj, Design error diagnosis and correction via test vector simulation. IEEE Trans. Comput. Aided Design Circuits Syst. **18**(12), 1803–1816 (1999)
87. C.S. Wallace, A suggestion for a fast multiplier. IEEE Trans. Electron. Comput. **EC-13**(1), 14–17 (1964)
88. U. Wilhelm, S. Ebel, A. Weitzel, Functional safety of driver assistance systems and ISO 26262, in *Handbook of Driver Assistance Systems: Basic Information, Components and Systems for Active Safety and Comfort* (Springer, Berlin, 2016), pp. 109–131
89. C. Wolf, Yosys Open SYnthesis Suite (2022). https://github.com/YosysHQ/yosys

90. W. Yang, L. Wang, A. Mishchenko, Lazy man's logic synthesis, in *International Conference on Computer-Aided Design* (2012), pp. 597–604
91. A. Yasin, T. Su, S. Pillement, M.J. Ciesielski, Formal verification of integer dividers: division by a constant, in *IEEE Annual Symposium on VLSI* (2019), pp. 76–81
92. A. Yasin, T. Su, S. Pillement, M.J. Ciesielski, Functional verification of hardware dividers using algebraic model, in *VLSI of System-on-Chip* (2019), pp. 257–262
93. C. Yu, W. Brown, D. Liu, A. Rossi, M. Ciesielski, Formal verification of arithmetic circuits by function extraction. IEEE Trans. Comput. Aided Design Circuits Syst. **35**(12), 2131–2142 (2016)
94. C. Yu, M. Ciesielski, A. Mishchenko, Fast algebraic rewriting based on and-inverter graphs. IEEE Trans. Comput. Aided Design Circuits Syst. **37**(9), 1907–1911 (2017)
95. R. Zimmermann, *Binary Adder Architectures for Cell-Based VLSI and Their Synthesis*. Ph.D. Thesis, Swiss Federal Institute of Technology, 1997
96. D. Zuras, W.H. McAllister, Balanced delay trees and combinatorial division in VLSI. IEEE J. Solid-State Circuits **21**(5), 814–819 (1986)

Index

A. Mahzoon et al., *Formal Verification of Structurally Complex Multipliers*,
https://doi.org/10.1007/978-3-031-24571-8

Printed in the United States
by Baker & Taylor Publisher Services